D0961460

The Case Against Sugar

This Large Print Book carries the
Seal of Approval of N.A.V.H.

THE CASE AGAINST SUGAR

GARY TAUBES

THORNDIKE PRESS
A part of Gale, Cengage Learning

Farmington Hills, Mich • San Francisco • New York • Waterville, Maine
Meriden, Conn • Mason, Ohio • Chicago

Portions of Chapter 8 originally in *Mother Jones*, (November/December 2012), as "Sweet Little Lies," coauthored by Gary Taubes and Cristin Kearns Couzens.

Thorndike Press, a part of Gale, Cengage Learning.

Thorndike Press® Large Print Lifestyles.

The text of this Large Print edition is unabridged.

Other aspects of the book may vary from the original edition.

Set in 16 pt. Plantin.

LIBRARY OF CONGRESS CATALOGING-IN-PUBLICATION DATA

Names: Taubes, Gary, author.
Title: The case against sugar / by Gary Taubes.
Description: Large print edition. | Waterville, Maine : Thorndike Press, a part of Gale, Cengage Learning, 2017. | Series: Thorndike Press large print lifestyles | Includes bibliographical references.
Identifiers: LCCN 2016057314| ISBN 9781410498953 (hardback) | ISBN 1410498956 (hardcover)
Subjects: LCSH: Sugar-free diet—Case studies. | Sugar—Physiological effect—Popular works. | Nutritionally induced diseases—Popular works. | Large type books. | BISAC: SCIENCE / Chemistry / General. | HEALTH & FITNESS / Nutrition.
Classification: LCC RM237.85 .T38 2017 | DDC 613.2/8332—dc23
LC record available at https://lccn.loc.gov/2016057314

Published in 2017 by arrangement with Alfred A. Knopf, a division of Penguin Random House LLC

Printed in the United States of America
1 2 3 4 5 6 7 21 20 19 18 17

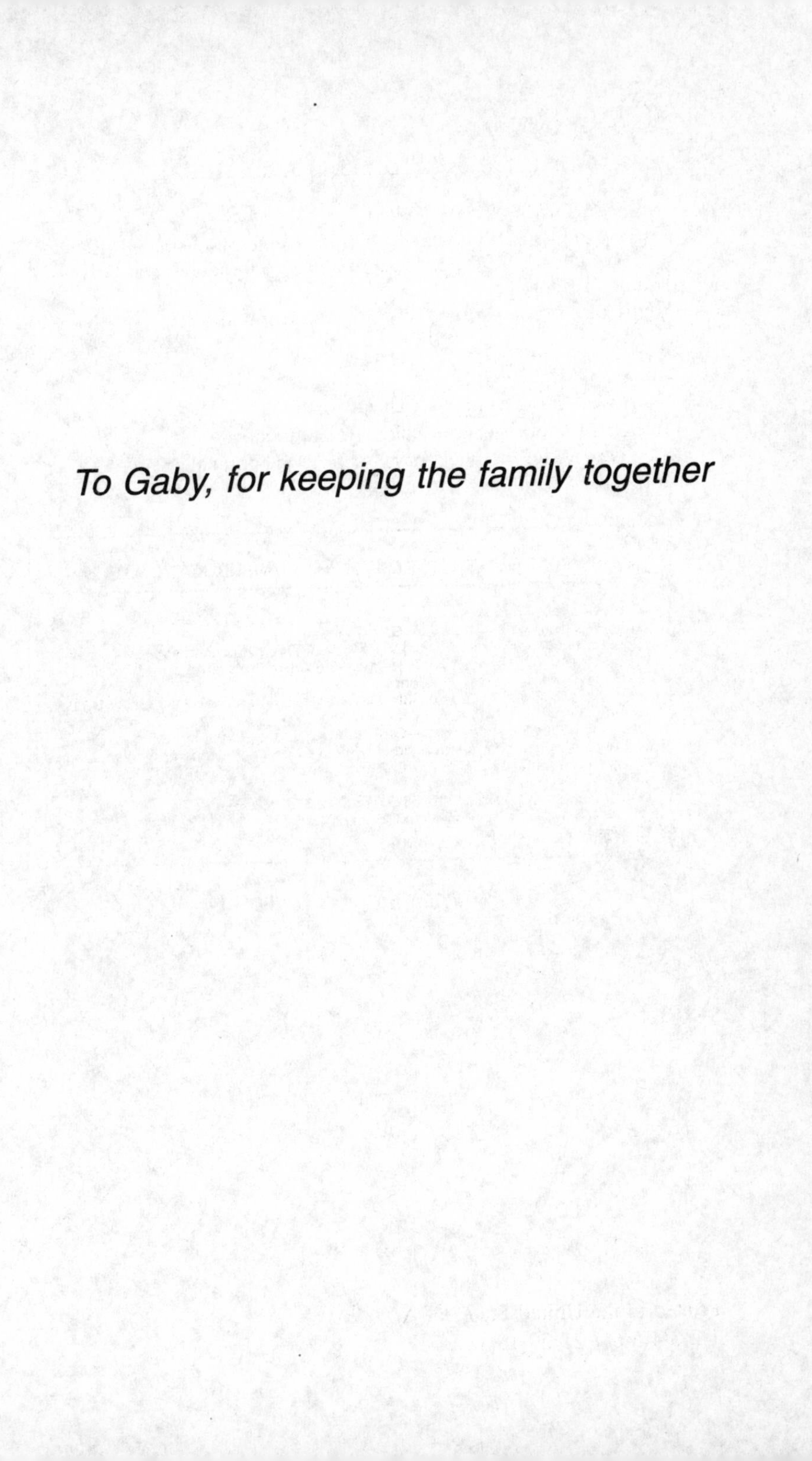

To Gaby, for keeping the family together

We are, beyond question, the greatest sugar-consumers in the world, and many of our diseases may be attributed to too free a use of sweet food.

The New York Times, May 22, 1857

I am not prepared to look back at my time here in this Parliament, doing this job, and say to my children's generation: I'm sorry, we knew there was a problem with sugary drinks, we knew it caused disease, but we ducked the difficult decisions and we did nothing.

GEORGE OSBORNE, U.K. chancellor of the exchequer, announcing a tax on sugary beverages, March 16, 2016

CONTENTS

AUTHOR'S NOTE

The purpose of this book is to present the case against sugar — both sucrose and high-fructose corn syrup — as the principal cause of the chronic diseases that are most likely to kill us, or at least accelerate our demise, in the twenty-first century. Its goal is to explain why these sugars are the most likely suspects, and how we arrived at the current situation: a third of all adults are obese, two-thirds overweight, almost one in seven is diabetic, and one in four to five will die of cancer; yet the prime suspects for the dietary trigger of these conditions have been, until the last decade, treated as little worse than a source of harmless pleasure.

If this were a criminal case, *The Case Against Sugar* would be the argument for the prosecution.

Introduction: Why Diabetes?

> Mary H— an unmarried woman, twenty-six years of age, came to the Out Patient Department of the Massachusetts General Hospital on August 2, 1893. She said her mouth was dry, that she was "drinking water all the time" and was compelled to rise three to four times each night to pass her urine. She felt "weak and tired." Her appetite was variable; the bowels constipated and she had a dizzy headache. Belching of gas, a tight feeling in the abdomen, and a "burning" in the stomach followed her meals. She was short of breath.
>
> ELLIOTT JOSLIN'S diabetes "case no. 1," as recorded in the case notes of his clinic

Elliott Joslin was a medical student at Harvard in the summer of 1893, working as a clinical clerk at Massachusetts General Hospital, when he documented his first consultation with a diabetic patient. He was still a good three decades removed from becoming the most influential diabetes specialist of the twentieth century. The patient was Mary Higgins, a young immigrant who had arrived from Ireland five years previously and had been working as a domestic in a Boston suburb. She had "a severe form of diabetes mellitus," Joslin noted, and her kidneys were already "succumbing to the strain put upon them" by the disease.

Joslin's interest in diabetes dated to his undergraduate days at Yale, but it may have been Higgins who catalyzed his obsession. Over the next five years, Joslin and Reginald Fitz, a renowned Harvard pathologist, would comb through the "hundreds of volumes" of handwritten case notes of the Massachusetts General Hospital, looking for information that might shed light on the cause of the disease and perhaps suggest how to treat it. Joslin would travel twice to Europe, visiting medical centers in Germany and Austria, to learn from the most influential diabetes experts of the era.

In 1898, the same year Joslin established his private practice to specialize in the treatment of diabetics, he and Fitz presented their analysis of the Mass General case notes at the annual meeting of the American Medical Association in Denver. They had examined the record of every patient treated at the hospital since 1824. What they saw, although they didn't recognize it at the time, was the beginning of an epidemic.

Among the forty-eight thousand patients treated in that time period, a year shy of three-quarters of a century, a total of 172 had been diagnosed with diabetes. These patients represented only 0.3 percent of all cases at Mass General, but Joslin and Fitz detected a clear trend in the admissions: the number of patients with diabetes and the percentage of patients with diabetes had both been increasing steadily. As many diabetics were admitted to Mass General in the thirteen years after 1885 as in the sixty-one years prior. Joslin and Fitz considered several explanations, but they rejected the possibility that the disease itself was becoming more common. Instead, they attributed the increase in diabetic patients to a "wholesome tendency of diabetics to place themselves under careful medical supervision." It wasn't that more Bostonians were succumbing to diabetes year to year, they said, but that a greater proportion of those who did were taking themselves off to the hospital for treatment.

By January 1921, when Joslin published an article about his clinical experience with diabetes for *The Journal of the American Medical Association,* his opinion had changed considerably. He was no longer talking about the wholesome tendencies of diabetics to seek medical help, but was using the word "epidemic" to describe what he was witnessing. "On the broad street of a certain peaceful New England village there once stood three houses side by side," he wrote, apparently talking about his hometown of Oxford, Massachusetts. "Into these three houses moved in succession four women and three men — heads of families — and of this number all but one subsequently succumbed to diabetes."

Joslin suggested that had these deaths been caused by an infectious disease — scarlet fever, perhaps, or typhoid, or tuberculosis — the local and state health departments would have mobilized investigative teams to establish the vectors of the disease and prevent further spread. "Consider the measures," he wrote, "that would have been adopted to discover the source of the outbreak and to prevent a recurrence." Because diabetes was a chronic disease, not an infectious one, and because the deaths occurred over years and not in the span of a few weeks or months, they passed unnoticed. "Even the insurance companies," Joslin wrote, "failed to grasp their significance."

We've grown accustomed, if not inured, to reading about the ongoing epidemic of obesity. Fifty years ago, one in eight American adults was obese; today the number is greater than one in three. The World Health Organization reports that obesity rates have doubled worldwide since 1980; in 2014, more than half a billion adults on the planet were obese, and more than forty million children under the age of five were overweight or obese. Without doubt we've been getting fatter, a trend that can be traced back in the United States to the nineteenth century, but the epidemic of diabetes is a more intriguing, more telling phenomenon.

Diabetes was not a new diagnosis at the tail end of the nineteenth century when Joslin did his first accounting, rare as the disease might have been then. As far back as the sixth century B.C., Sushruta, a Hindu physician, had described the characteristic sweet urine of diabetes mellitus, and noted that it was most common in the overweight and the gluttonous. By the first century A.D., the disease may have already been known as "diabetes" — a Greek term meaning "siphon" or "flowing through" — when Aretaeus of Cappodocia described its ultimate course if allowed to proceed untreated: "The patient does not survive long when it is completely established, for the marasmus [emaciation] produced is rapid, and death speedy. Life too is odious and painful, the thirst is ungovernable, and the copious potations are more than equaled by the profuse urinary discharge. . . . If he stop for a very brief period, and leave off drinking, the mouth becomes parched, the body dry; the bowels seem on fire, he is wretched and uneasy, and soon dies, tormented with burning thirst."

Through the mid-nineteenth century, diabetes remained a rare affliction, to be discussed in medical texts and journal articles but rarely seen by physicians in their practices. As late as 1797, the British army surgeon John Rollo could publish "An Account of Two Cases of the Diabetes Mellitus," a seminal paper in the history of the disease, and report that he had seen these cases nineteen years apart despite, as Rollo wrote, spending the intervening years "observ[ing] an extensive range of disease in America, the West Indies, and in England." If the mortality records from Philadelphia in the early nineteenth century are any indication, the city's residents were as likely to die from diabetes, or at least to have diabetes attributed as the cause of their death, as they were to be murdered or to die from anthrax, hysteria, starvation, or lethargy.*

In 1890, Robert Saundby, a former president of the Edinburgh Royal Medical Society, presented a series of lectures on diabetes to the Royal College of Physicians in London in which he estimated that less than one in every fifty thousand died from the disease. Diabetes, said Saundby, is "one of those rarer diseases" that can only be studied by physicians who live in "great cent[er]s of population and have the extensive practice of a large hospital from which to draw their cases." Saundby did note, though, that the mortality rate from diabetes was rising throughout England, in Paris, and even in New York. (At the same time, one Los Angeles physician, according to Saundby, reported "in seven years' practice

* At Massachusetts General Hospital, the very same handwritten medical records that Joslin would later analyze reveal that for twenty of the forty-five years between 1824 and 1869 there was not a single case of diabetes. In none of these years were there more than three cases.

he had not met with a single case.") "The truth," Saundby said, "is that diabetes is getting to be a common disease in certain classes, especially the wealthier commercial classes."

William Osler, the legendary Canadian physician often described as the "father of modern medicine," also documented both the rarity and the rising tide of diabetes in the numerous editions of his seminal textbook, *The Principles and Practice of Medicine.* Osler joined the staff at Johns Hopkins Hospital in Baltimore when the institution opened in 1889. In the first edition of his textbook, published three years later, Osler reported that, of the thirty-five thousand patients under treatment at the hospital since its inception, only ten had been diagnosed with diabetes. In the next eight years, 156 cases were diagnosed. Mortality statistics, wrote Osler, suggested an exponential increase in those reportedly dying from the disease — nearly doubling between 1870 and 1890 and then more than doubling again by 1900.

By the late 1920s, Joslin's epidemic of diabetes had become the subject of newspaper and magazine articles, while researchers in the United States and Europe were working to quantify accurately the preva-

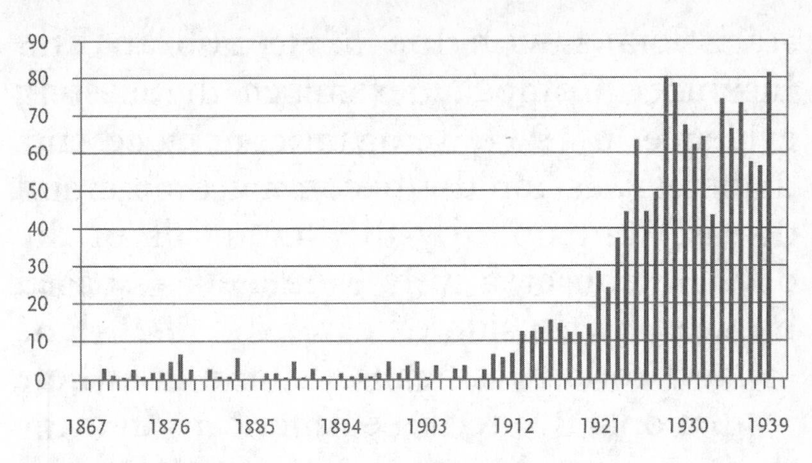

THE BEGINNINGS OF AN EPIDEMIC? **Diabetes admissions, Pennsylvania Hospital, Philadelphia**

lence of the disease, in a way that might allow meaningful comparisons to be drawn from year to year and decade to decade. In Copenhagen, for instance, the number of diabetics treated in the city's hospitals increased from ten in 1890 to 608 in 1924 — a sixty-fold increase. When the New York City health commissioner Haven Emerson and his colleague Louise Larimore published an analysis of diabetes mortality statistics in 1924, they reported a 400 percent increase in some American cities since 1900 — almost 1,500 percent since the Civil War.

Despite all this, the disease remained a relatively rare one. When Joslin, working with Louis Dublin and Herbert Marks, both statisticians with the Metropolitan Life Insurance Company, examined the existing evidence in 1934, he again concluded that diabetes was rapidly becoming a common disease, but only by the standards of the day. He conservatively estimated — based on what he considered careful studies done in New York, Massachusetts, and elsewhere — that only two to three Americans in every thousand had diabetes.

Times have certainly changed. In 2012, the latest year for which the Centers for Disease Control (CDC) have provided estimates, one in every seven to eight adults in this country had diabetes — 12 to 14 percent, depending on the criteria used to diagnose it. Another *30 percent* are predicted to get diabetes at some point during their lives. Almost two million Americans were diagnosed with diabetes in 2012 — one case every fifteen to sixteen seconds. Among U.S. military veterans, one in every four patients admitted to VA hospitals suffers from diabetes.

The great proportion of this tidal wave of diabetics — perhaps 95 percent — have what is now known as type 2 diabetes, the form of the disease, as Sushruta would have said over two thousand years ago, that associates with overweight and obesity. A small proportion have type 1, typically children. This is the acute form of the disease, and it kills, if untreated, far more quickly.* Both type 1 and type 2 diabetes have been increasing in prevalence for the past 150 years; in both, the increase has been dramatic.

Those afflicted with diabetes will die at greatly increased rates from heart disease or stroke, from kidney disease — the disease is now considered the cause of more than 40 percent of cases of kidney failure — and diabetic coma. Without appropriate treatment (and occasionally even with), their eyesight will deteriorate (often a first symptom); they'll suffer nerve damage; their teeth will decay and fall out; they'll get foot ulcers and gangrene; and they'll lose limbs to amputation. Six in every ten lower-limb amputations in adults are due to diabetes — some seventy-three thousand of them in 2010 alone. A dozen *classes* of drugs are now available to treat the disease, and the market for diabetic drugs and devices in the

* Because type 2 diabetes is so much more common, when I refer to diabetes in this book I will be referring to the type 2 form or both type 2 and type 1 together, unless specified otherwise.

United States alone is over thirty billion dollars yearly. Drugstore chains now offer free tests to customers to check levels of blood sugar, hoping to sell home-testing kits to those whose blood sugar might happen to show up borderline or high.

The obvious questions are: Why have things changed so? How did we get here? What forces of nature or environment or lifestyle have led to diabetes in one out of every eleven Americans, children and adults together?

One way to avoid answering this question is to assume that historical trends in diabetes prevalence constitute unreliable evidence. Who knows what was really going on fifty or a hundred years ago? And, indeed, it's surprisingly difficult to quantify with any confidence the changing prevalence of a chronic disease in a population. Such issues as the criteria by which it's diagnosed, how much attention physicians, the public, and the media pay to it, the availability of treatment and how well those treatments work, the longevity of the population, and whether the disease is more common with age will all confound any authoritative attempts to establish reliably how the actual occurrence of a chronic disease has changed with time. It's a very good bet, though, that had one in eleven Americans been afflicted with diabetes in the nineteenth century, the hospital inpatient records of those eras would have looked dramatically different, as would the number of deaths attributed to diabetes. As Saundby wrote in 1901, "Diabetes is in all cases a grave disease. . . . Life seems to hang by a thread, a thread often cut by a very trifling accident."

For the past century, the observation that diabetes is increasing in the population — transitioning from a rare disease to a common one and now to a scourge — has remained a constant theme in the medical literature. In 1940, Russell Wilder, the leading diabetologist at the Mayo Clinic, reported that diabetes admissions had been increasing steadily at the clinic for the previous twenty years. "The incidence of diabetic morbidity is unknown," he wrote, "but the indications that it is increasing are very clear." Ten years later, Joslin himself referred to the "appalling increase in diabetes," which he now considered an inescapable fact of life. In 1978, Kelly West, the leading American authority on diabetes epidemiology — the study of how diseases move through populations — suggested that diabetes had already killed more people in the twentieth century than all wars combined. "Diabetes mellitus has become one of the most important of human problems," he wrote, calling it "a significant cause of disease and death in all countries and all major races."

Epidemic increases in the occurrence of diabetes, as West suggested, were not a localized phenomenon. Diabetes was virtually unknown or at least undiagnosed in China, for instance, at the turn of the twentieth century. One British physician reported seeing only one case of the disease among twenty-four thousand outpatients in Nanking, although "all drawn from the lower classes of society." Another reported only two cases among the twelve thousand inpatients treated in his hospital. In the 1980s, the prevalence of diabetes in the Chinese population at large was still estimated to be approximately 1 percent. The latest estimates are that 11.6 percent of the adult population is diabetic — one in nine, more than 110 million Chinese in total. Almost half a *billion* Chinese are believed to be pre-diabetic.

The prevalence of both diabetes and prediabetes was considered vanishingly small among Inuit in Greenland, Canada, and Alaska through the 1960s — "Eight Alaskan Eskimos are now known to have diabetes," reported one article in the *Journal of the American Medical Association* in 1967. By the 1970s, diabetes was still rare, but researchers were now documenting the increasing appearance of a pre-diabetic condition, glucose intolerance. In recent studies, diabetes rates in the Inuit are now at 9 percent — one in every eleven individuals — similar to the levels in Canada and the United States as a whole.

The same epidemic patterns have been observed in Native American tribes (particularly the Pima population in Arizona, as we'll discuss later) and in the First Nations People of Canada. In many of these populations, one out of every two adults now has diabetes. In some cases — the Ojibwa Cree people of Sandy Lake in northern Ontario, for instance — diabetes, if it existed, was undiagnosed in the population as late as the 1960s. In 1974, when Kelly West examined the available data on diabetes in Native American populations, he concluded that the disease had been rare to nonexistent prior to the 1940s — both civilian and military physicians had carried out health surveys — and yet, by the mid-1960s, research, including his own, was documenting previously unafflicted populations in which one in four adults was diabetic.

(When researchers charted the number of cases diagnosed each year in the Navajo from the 1950s through the 1980s, the resulting graph looked almost identical to that on page 8 from Pennsylvania Hospital in Philadelphia a century earlier.) Similar patterns have been observed in Polynesians, Micronesians, and Melanesians in the South Pacific; in aboriginal populations in Australia; in Maoris in New Zealand; and in populations throughout the Middle East, Asia, and Africa. In fact, anywhere populations begin eating Western diets and living Western lifestyles — whenever and wherever they're acculturated or urbanized, as West noted in 1978 — diabetes epidemics follow.

So what happened? What's happening? Something changed dramatically in our diets, our lifestyle, or our environment to trigger these unprecedented epidemics of diabetes; but what? As Joslin observed under similar circumstances at a far earlier stage in this epidemic, had this been an infectious disease, the relevant boards of health, the insurance agencies, the newspapers, the country as a whole, would be demanding answers. The CDC and the World Health Organization would have established panels of expert investigators to pry into every crevice of our assumptions about the cause of this disease to see where we might have misunderstood its etiology. Such is not the case.

Prior to the 1970s, public-health authorities and clinicians commenting on the rising tide of diabetes in the populations they studied frequently suggested what to them seemed like the prime suspect — sugar consumption. Here was a disease of carbohydrate metabolism that was becoming increasingly common as populations began consuming sugar — a kind of carbohydrate — at levels that were virtually unimaginable a century before; in some cases, just twenty or thirty years before.

As sugar consumption exploded in the United States and the United Kingdom with the industrial revolution; with the birth of the confectionary, cereal, and soft-drink industries; and with the increasing availability of chocolate bars and ice-cream treats, so did diabetes begin its inexorable climb. When sugar and sugar-rich products spread around the globe, so did diabetes. When peasant farmers throughout Africa, India, Asia, and Central and South America migrated to towns and cities to become wage earners, and changed their dietary habits accordingly — no longer eating locally grown cereals, starches, and fruits, but instead buying sugary drinks and sugar-laden treats in shops and markets — diabetes made its inevitable appearance. As Kelly West said about the emerging epidemics of diabetes in Native American populations in 1974, "Some had been nomadic hunters and meat eaters . . . while others had derived a substantial majority of their calories from fats. . . . Sugar consumption has been increasing in most, if not all, of the United States tribes in whom diabetes rates have recently increased precipitously. This same association has been observed in Eskimos of Alaska, Canada, and Greenland as well as in Polynesians."

And on those very rare occasions when sugar consumption declined — as it did, for instance, during World War I, because of government rationing and sugar shortages — diabetes mortality invariably declined with it. "Rises and falls in sugar consumption," wrote Haven Emerson and Louise Larimore in 1924, "are followed with fair regularity . . . by similar rises and falls in the death rates from diabetes."

In 1974, when the sugar industry hired pollsters to survey physicians for their attitudes toward sugar, most of those physicians said they thought sugar consumption accelerated the onset of diabetes. (One advertising executive, later asked if his children ate a particularly sugar-rich cereal for which he had modeled the ad campaign on Snoopy and the Red Baron, admitted that they never did: "You need an insulin shot if you eat a bowl of that," he said.) In 1973, Jean Mayer of the Harvard School of Public Health, probably the most influential nutritionist of the era, was suggesting that sugar "plays an etiological role in those individuals who are genetically susceptible to the disease." Such a statement, of course, raises the obvious question of whether anyone ever gets the disease who *isn't* genetically susceptible (with the rare exceptions of those individuals who sustain injuries or tumors that affect pancreatic function). Nonetheless, at scientific meetings on sugar and other sweeteners, researchers and clinicians would debate whether or not sugar caused diabetes or only helped it along in those somehow predisposed.

By the late 1970s, though, sugar had mostly vanished from the discussion. Dietary fat had been implicated as a cause of heart disease. Nutritionists and public-health authorities responded by rejecting the idea that sugar could be responsible for the diseases that associated with heart disease, which included both obesity and diabetes.

Researchers had also come to embrace a pair of related assumptions that were poorly tested and might or might not be true. The first is that type 2 diabetes is caused by obesity, because the two diseases are so closely associated, both in populations and in individuals, and obesity typically appears first (although more than one in every ten individuals diagnosed with type 2 diabetes is neither obese nor overweight). The second assumption, as the World Health Organization puts it, is: "The fundamental cause of obesity and overweight is an energy imbalance between calories consumed and calories expended." "The only trouble with the American diet," as Fred Stare, the founder and head of the nutrition department at Harvard University, said in 1976 on national television, is that "we eat too damn much." The overeating was accompanied by a decrease in physical activity, attributed to changing modes of transportation and the mechanization of labor.

Public-health authorities have considered no investigations necessary to explain the obesity and diabetes epidemics, because they have assumed that the cause is obvious. Attempts to prevent diabetes in the

United States, Europe, and Asia, and among populations worldwide, are almost invariably aimed at getting these populations to eat smaller portions and fewer calories, perhaps to avoid "fatty foods," as particularly dense sources of calories, and to increase their physical activity.

Meanwhile, the latest surge in this epidemic of diabetes in the United States — an 800 percent increase from 1960 to the present day, according to the Centers for Disease Control — coincides with a significant rise in the consumption of sugar. Or, rather, it coincides with a surge in the consumption of *sugars,* or what the FDA calls "caloric sweeteners" — sucrose, from sugarcane or beets, and high-fructose corn syrup, HFCS, a relatively new invention.

After ignoring or downplaying the role of sugars and sweets for a quarter-century, many authorities now argue that these are, indeed, a major cause of obesity and diabetes and that they should be taxed heavily or regulated. The authorities still do so, however, not because they believe sugar causes disease but, rather, because they believe sugar represents "empty calories" that we eat in excess because they taste so good. By this logic, since refined sugar and high-fructose corn syrup don't contain any protein, vitamins, minerals, antioxidants, or fiber, they either displace other, more nutritious elements of our diet, or simply add extra, unneeded calories to make us fatter. The Department of Agriculture, for instance (in its recent "Dietary Guidelines for Americans"), the World Health Organization, and the American Heart Association, among other organizations, advise a reduction in sugar consumption for these reasons primarily.

The empty-calories argument is particularly convenient for the food industry, which would understandably prefer not to see a key constituent of its products — all too often, *the* key constituent — damned as toxic. The sugar industry played a key role in the general exoneration of sugar that took place in the 1970s, as I'll explain later. Health organizations, including the American Diabetes Association and the American Heart Association, have also found the argument convenient, having spent the last fifty years blaming dietary fat for our ills while letting sugar off the hook.

The empty-calories logic allows companies that sell sugar-rich products, or products in which all the calories come from these sugars, to claim that they, too, are fighting the good fight. They can profess and perhaps believe that they are fighting the scourge of childhood obesity and diabetes — that they are part of the solution, not the problem — by working to educate children on how to eat less, be satisfied with smaller portions, and exercise more, just as Coca-Cola, PepsiCo, Mars, Nestlé, Hershey's, and a few dozen other companies did in 2009 when they joined up with the Grocery Manufacturers Association, the American Dietetic Association (now the Academy of Nutrition and Dietetics), and the Girl Scouts of the USA to found the Healthy Weight Commitment Foundation. Embracing the notion of empty calories is politically expedient as well. Any politician running for public office is unlikely to benefit from alienating major constituents of the food industry, particularly companies with powerful lobbies, such as the sugar and beverage industries. "This is not about demonizing any industry," as Michelle Obama said in 2010 about "Let's Move," her much-publicized program to combat childhood obesity.

This book makes a different argument: that sugars like sucrose and high-fructose corn syrup are fundamental causes of diabetes and obesity, using the same simple concept of causality that we employ when we say smoking cigarettes causes lung cancer. It's not because we eat too much of these sugars — although that is implied merely by the terms "overconsumption" and "overeating" — but because they have unique physiological, metabolic, and endocrinological (i.e., hormonal) effects in the human body that directly trigger these disorders. This argument is championed most prominently by the University of California, San Francisco, pediatric endocrinologist Robert Lustig. These sugars are not short-term toxins that operate over days and weeks, by this logic, but ones that do their damage over years and decades, and perhaps even from generation to generation. In other words, mothers will pass the problem down to their children, not through how and what they feed them (although that plays a role), but through what they eat themselves and how that changes the environment in the womb in which the children develop.

Individuals who get diabetes — the ones in any population who are apparently susceptible, who are genetically predisposed — would never have been stricken if they (and maybe their mothers and their mothers' mothers) lived in a world without sugar, or at least in a world with a lot less of it than the one in which we have lived for the past

100 to 150 years. These sugars are what an evolutionary biologist might call the environmental or dietary trigger of the disease: the requisite ingredient that triggers the genetic predisposition and turns an otherwise healthy diet into a harmful one. Add such sugars in sufficient quantity to the diet of any population, no matter what proportion of plants to animals they eat — as Kelly West suggested in 1974 about Native American populations — and the result eventually is an epidemic of diabetes, and obesity as well. If this is true, then to make headway against these disorders — to prevent future cases of obesity and diabetes from manifesting themselves, and to reverse the epidemics that are now ongoing — we must show these sugars and the businesses that sell them for what they truly are.

The implications of the case against sugar go far beyond diabetes. Those who are obese or diabetic are also more likely to have fatty liver disease, and this, too, is now epidemic in Westernized populations. The National Institutes of Health estimate that as many as one in four Americans now have the disease, unrelated to alcohol consumption. If untreated, it can progress to cirrhosis of the liver and eventually the need for a liver transplant. Those who are obese and diabetic also tend to be hypertensive; they have a higher risk of heart disease, cancer, and stroke, and possibly dementia and even Alzheimer's disease as well.

These chronic diseases — the diseases that ultimately kill us in modern Western societies — tend to cluster together in both populations and individual patients. Diabetes, heart disease, cancer, stroke, and Alzheimer's account for five of the top ten causes of death in the U.S. A conservative estimate is that they cost the medical system and our society, in lost work and productivity, one trillion dollars a year.

Together they're often referred to as diseases of Western lifestyles, or diseases of Westernization. This cluster has led cancer researchers to suggest that obesity is a cause of cancer. It has led some Alzheimer's researchers to refer to Alzheimer's as type 3 diabetes.

All of these diseases have now been linked to a condition known as "insulin resistance," a phenomenon we will examine in depth. Insulin resistance is the fundamental defect present in type 2 diabetes and perhaps obesity as well. So it's a reasonable possibility that the same thing that causes one of these diseases — type 2 diabetes in particular — causes all of them. It's what scientists would call the null hypothesis, a starting point for research, discussion, and studies. If sugar and high-fructose corn syrup are the cause of obesity, diabetes, and insulin resistance, then they're also the most likely dietary trigger of these other diseases. Put simply: without these sugars in our diets, the cluster of related illnesses would be far less common than it is today; likewise other disorders that associate with these illnesses, among them polycystic ovary syndrome (PCOS), rheumatoid arthritis, gout, varicose veins, asthma, and inflammatory bowel disease.

If this were a criminal investigation, the detectives assigned to the case would start from the assumption that there was one prime suspect, one likely perpetrator, because the crimes (all the aforementioned diseases) are so closely related. They would only embrace the possibility that there were multiple perpetrators when the single-suspect hypothesis was proved insufficient to explain all the evidence. Scientists know this essential concept as Occam's Razor. When Isaac Newton said, "We are to admit no more causes of natural things than such as are both true and sufficient to explain their appearances," he was saying the same thing that Albert Einstein, three centuries later, said (or was paraphrased as saying): "Everything should be made as simple as possible, but no simpler." We should begin with the simplest possible hypothesis, and only if that can't explain what we observe should we consider more complicated explanations — in this case, multiple causes.

This is not, however, how medical researchers and public-health authorities have come to think about these disorders. Despite their faith in the notion that obesity causes or accelerates diabetes and that therefore (in what I will argue is a mistaken assumption) both are diseases of overconsumption and sedentary behavior, they will also defend their failure to curb the ongoing epidemics of these diseases on the basis that these are "multifactorial, complex disorders" or "multidimensional diseases." By this they mean that so many factors are involved in the genesis and progression of these diseases — including genetics for sure, epigenetics (the modification of how genes are turned on and off in cells), how much we eat and exercise, perhaps how well we sleep, toxins in the environment, pharmaceuticals, possibly viruses, the effect of antibiotic use on the bacteria in our guts (dysbiosis, as it's now commonly called, or microbial imbalance) — that to identify one ultimate trigger, or one critical component of our modern diets, is to be naïve.

The counterargument is simple: Lung cancer is also assuredly a multifactorial, complex disease. Most smokers will never get lung cancer, and at least a tenth of all cases of lung cancer are unrelated to smoking cigarettes, and yet it's widely accepted — for very good reasons — that smoking is *the* primary cause. Whether or not obesity and diabetes and their associated diseases are multifactorial, complex disorders, something has to explain their connection with modern Western diets and lifestyles and the epidemics that are both ongoing and almost ubiquitous worldwide. What is it? We are clearly doing something different from what we did fifty years ago, or 150 years ago, and our bodies and health reflect it. Why?

The goal of this book is to clarify the arguments against sugar, correct some of the misconceptions and preconceptions that have dogged the debate for the hundreds of years during which it's been ongoing, and provide the perspective and context needed to make reasonable decisions on sugar as individuals and as a society. People are dying today, literally every second, from diseases that seemed virtually nonexistent in populations that didn't eat modern Western diets or live modern Western lifestyles. Something is killing them prematurely. This book will document the case against sugar as the prime culprit.

In my two previous books on health and nutrition, I discussed the evidence implicating all highly processed and easily digestible carbohydrates in general — grains and starchy vegetables — as well as sugar and high-fructose corn syrup. I suggested that there was something unique about those sugars that then made the other carbohydrate-rich foods a problem as well. So the treatment of the conditions they caused — particularly obesity and diabetes — often required restricting some or all of these carbohydrates, not just sugar.

In this book, the focus is specifically on the role of sugar in our diet, and the likely possibility that the difference between a healthy diet and one that causes obesity, diabetes, heart disease, cancer, and other associated diseases begins with the sugar content. If this is true, it implies that populations or individuals can be at the very least reasonably healthy living on carbohydrate-rich diets, even grain-rich diets, as long as they consume relatively little sugar. As sugar consumption rises and people ingest it over decades, and across generations, it causes insulin resistance and triggers the progression to obesity, diabetes, and the diseases that associate with them. Once this process starts, easily digestible, carbohydrate-rich foods aid and abet it. If the argument is correct, the first necessary step in preventing or avoiding these diseases is to remove the sugars from our diets.

This argument also serves to censure the last century of advice on obesity, diabetes, and nutrition, notwithstanding the best intentions of those who gave it. Despite a century's worth of evidence implicating sugar as the cause of insulin resistance and diabetes and many, perhaps all, of the diseases that associate with them, the researchers working in these fields, and the health organizations funding this research, chose to ignore it or reject it. Invariably, they did so on the basis of ill-founded assumptions and preconceptions about what other factors might be responsible — dietary fat, or the simplistic idea that eating too many calories of any kind makes us fat. Here I'll be discussing the science as much as the errors in judgment that were made during this time. It's one thing to claim that sugar is uniquely toxic — perhaps having prematurely killed more people than cigarettes or "all wars combined," as Kelly West said about diabetes itself — but to do so convincingly we have to understand why this conclusion has not been common wisdom.

In the process, I'll be looking at the key scientific issues with a decidedly historical perspective. History is critical to understanding science and how it progresses. In many scientific disciplines — physics, for example — the science is taught with the history attached. Students learn not only what is believed to be true and which conjectures have fallen by the wayside, but on the basis of what experiments and what evidence, and by whose authority and ingenuity. The names of the physicists responsible for the advances in understanding — Newton, Einstein, Maxwell (for his equations of electromagnetism); Heisenberg, Planck, and Schrödinger, among others, for their work in understanding the quantum nature of the universe; and many more — are as well known as many historical figures in politics and other fields. Medicine today, though, as with related fields such as nutrition, is taught mostly untethered from its history. Students are taught what to believe but not always the evidence on which these beliefs are based, and so oftentimes the beliefs cannot be questioned. And medical students are not taught, as physics students typically are, to question everything that has not demonstrably survived the trial-by-fire process of rigorous, methodical testing. Students of any science need to know why they are being asked to believe a particular idea, or why not, and on what grounds. Without the history of the idea, there's no way to tell and, by implication, no reason to ask.

This is why authorities on diabetes today will often argue that sugar does not cause diabetes but will do so based on little or no awareness of how that conclusion was ultimately reached and on what evidence. It's why the provenance of the idea that we get fat because we consume more calories than we expend is little known, even by those physicians and researchers who have been (or still are) its die-hard proponents. It's why the existence of a competing hypothesis of obesity as a hormonal disorder is little known, let alone that this hypothesis is capable of explaining the data and the observations in a way that the "energy balance" notion is incapable of doing.

In writing this book, I hope to continue to restore this history to the discussion of how our diets influence our weight and health, and to do so in the context of the vitally important question of sugar in the diet.

I want to clarify a few final points before we continue.

First, I'm going to concede in advance a key point that those who defend the role of sugar in our diet will invariably make. The sugar industry and purveyors of sugar-rich products are right when they say that it cannot be established definitively, with the science as it now stands, that sugar is uniquely harmful — a toxin that does its damage over decades. The evidence is not as clear with sugar as it is with tobacco. This isn't a failure of science but, rather, an issue of its limits.

With tobacco, researchers could compare smokers with nonsmokers and look for the difference in incidence of a single disease — lung cancer — that in nonsmokers, at least, is very rare. These studies were first done in the late 1940s, and the difference observed in these comparisons was so dramatic — heavy smokers had twenty to thirty times the risk of those who had never smoked — that it was effectively impossible to imagine any reasonable explanation other than cigarettes (not that the tobacco industry didn't try).

With sugar, the best researchers can do is compare individuals all of whom have consumed tremendous amounts of sugar, at least compared with the levels of consumption in nonindustrialized societies. If they compare sugar consumers with those who abstain, they're looking at individuals who have vastly different philosophies about how to lead a healthy life and so will differ in many meaningful ways other than just how much sugar they consume. They're also looking at differences in rates of what are now all-too-common diseases, although whether the diseases would be common in a world without sugar is the question. The study of sugar consumers versus nonconsumers entails issues and challenges that simply didn't exist in the study of cigarettes and lung cancer.

One way to tackle this problem is to compare populations that had no access to sugar, or very little, with those that had plenty — often the same populations twenty, fifty, or a hundred years later. Still, the difference in sugar consumption is just one of the many differences that might explain the differences in health status. It's possible to assemble a compelling argument with this method (just as a good prosecutor can create a compelling case from circumstantial evidence), but that is not sufficient to establish definitively what is causing the health effects we're seeing.

Whether we can assemble the kind of evidence that would stand up in a court of law and allow governments to regulate sugar, as they already do tobacco and alcohol, remains to be seen. But whether we have enough evidence and reasonable assumptions to convince ourselves to avoid sugar, to minimize its consumption, and to convince our children to do the same is a different question. That's the question this book will try to answer.

Second, I need to clarify what exactly we're talking about when we talk about sugar or sugars. This may seem obvious, but it certainly hasn't been in the past. The controversy over the health effects of sugar — proceeding, as it has, for hundreds of years — is littered with erroneous statements and conclusions that have driven thinking to the current day. Often, if not largely, it is because the individuals considered authorities on the subject often had no true understanding of what they were talking about, and thus no understanding of how different types of sugars — all carbohydrates — might have profoundly different effects on human health. This confusion still exists and still haunts some of the most influential reporting on diet and health, despite the multitudes of articles written on sugar and health in the past decade.

Biochemically, the term "sugar" refers to a group of carbohydrate molecules consisting, as the word "carbohydrate" implies, of atoms of carbon and hydrogen. The names of these carbohydrates all end in "-ose" — glucose, galactose, dextrose, fructose, lactose, sucrose, etc. All of these sugars will dissolve in water, and they all taste sweet to us, although to a greater or lesser extent. When physicians or researchers refer to "blood sugar," they're talking about glucose, because it constitutes virtually all of the sugar circulating in our blood.

The more common usage of "sugar" refers to sucrose, the white crystalline variety that we put in our coffee or tea or sprinkle on our morning cereal. Sucrose in turn is composed of equal parts glucose and fructose, the two smaller sugars (monosaccharides, in the chemical lingo) bonded together to make the larger one (a disaccharide). Fructose, found naturally in fruits and honey, is the sweetest of all these sugars, and it's the fructose that makes sucrose particularly sweet. Lately, researchers have been asking whether fructose is toxic, because it's the significant amount of fructose in sugar (sucrose) that differentiates it from other carbohydrate-rich foods, such as bread or potatoes, which break down upon digestion to mostly glucose alone. Because we never consume the fructose without the glucose, though, the appropriate question is whether sucrose, the combination of roughly equal parts fructose *and* glucose, is toxic, not one alone.

This would be confusing enough without the introduction in the 1970s of high-fructose corn syrup (HFCS), which replaced a significant part of the refined sugar (i.e., sucrose) consumed in the United States over the decade that followed. High-fructose corn syrup comes in different formulations; the most common one is known as HFCS-55, because it's 55 percent fructose and 45 percent glucose.* In sucrose, the ratio is 50-50. It was created, in fact, to replace sucrose inexpensively when used as the sweetener in soft drinks — specifically Coca-Cola — without any no-

* This ratio was called into question in a 2010 analysis claiming that fructose content in some popular sugary beverages was then as high as 65 percent.

ticeable difference in taste or sweetness.

The U.S. Department of Agriculture includes both sucrose and HFCS in the category of "caloric" or "nutritive" sweeteners, along with honey and maple syrup — both glucose-fructose combinations — differentiating them from artificial sweeteners such as saccharin, aspartame, and sucralose, which are effectively calorie-free. Public-health authorities often refer to sucrose and HFCS as "added sugars" to differentiate them from the component sugars that can be found naturally (in relatively small proportions) in fruits and vegetables.

Because the introduction of HFCS-55 roughly coincided with the beginning of the obesity epidemic in the United States, researchers and journalists later suggested that HFCS was the cause, implying that it was somehow distinct from sugar itself. HFCS was promptly demonized as a particularly pernicious aspect of the diet — "the flashpoint for everybody's distrust of processed foods," as the New York University nutritionist Marion Nestle has described it — and that's often still considered to be the case. This is why cans of Pepsi sweetened by sucrose rather than high-fructose corn syrup proudly proclaim that they contain "natural sugar." Newman's Own lemonade, sweetened with sucrose ("cane sugar," as the label says), proclaims prominently on the carton that it contains "no high fructose corn syrup." In 2010, the Corn Refiners Association petitioned the Food and Drug Administration to allow it to refer to high-fructose corn syrup as "corn sugar" on food labels, thus trying to avoid this demonization process. The sugar industry promptly sued them to prevent it from happening, at which point the Corn Refiners countersued. In 2012, the FDA denied the Corn Refiners' petition — sugar, the FDA said, "is a solid, dried, and crystallized sweetener" and HFCS is not — and so the latter is still clearly identifiable as both syrupy and derived from corn.

All of this controversy, however, though it may benefit the sugar (sucrose) industry in particular, serves only to obfuscate the key point: high-fructose corn syrup is not fructose, any more than sucrose is. (The reason for the appellation "high fructose" is that HFCS has a greater proportion of fructose to glucose than previous corn syrups, which date back to the nineteenth century and were never sweet enough to challenge the primacy of sucrose in foods and beverages.) Our bodies appear to respond the same way to both sucrose and HFCS. In a 2010 review of the relevant science, Luc Tappy, a researcher at the University of Lausanne in Switzerland, who is considered by biochemists who study fructose to be among the world's foremost authorities on the subject, said there was "not the single hint" that HFCS was more deleterious than other sources of sugar. The question I'll be addressing in this book is whether they are both benign, or both harmful — not whether one is worse than the other.

My usage of the words "sugar" or "sugars" throughout the text will depend on context. If I'm speaking about the present, when sucrose and high-fructose corn syrup are used to an equal extent, I'll use "sugar" to refer to both. If the context is prior to the introduction of high-fructose corn syrup in the late 1970s, then "sugar" will only mean sucrose, and I'll often qualify it by describing it as either beet sugar or cane sugar. If I'm referring to specific (monosaccharide) *sugars* — fructose, glucose, lactose, etc. — then that, too, will be clear from the context.

The last issue that requires clarification before we continue is that of how much of these sugars (i.e., caloric sweeteners) we actually consume or, for that matter, ever did. Through the 1970s, the per capita consumption numbers cited by government organizations, historians, and journalists — the numbers I typically use in this book — would have been for sugar "deliveries," as the Department of Agriculture now refers to them. This is the amount that industry makes available for consumer use. The formula is simple: domestic production plus imports minus exports, all divided by the current population. Governments acquire these numbers for tax, tariff, and other purposes, and they do it reasonably well. Hence, these numbers are (relatively) reliable, as are trends based on these numbers. We can assume, for instance, that when the USDA reports that 114 pounds of sugar and HFCS were delivered to retailers in 2014, that number can be meaningfully compared with the 153 pounds delivered in 1999, when deliveries (and, so we assume, consumption) peaked in the United States, and both can be compared with the few tens of pounds delivered per capita two hundred years ago.

Beginning in the 1980s, however, with a Food and Drug Administration report that we will discuss in chapter 8, authorities have often tried to estimate how much of this available sugar is actually consumed. After all, much gets thrown out with stale bakery products, for instance, or flat soda or the juice at the bottom of a cup or can. The authorities base these estimates primarily on surveys in which individuals are asked to recall what they ate and drank. This survey data is known to be exceedingly *un*reliable, which the USDA readily admits. ("Limitations on accurately measuring food loss," it says, "suggest that actual loss rates may differ from the assumptions used.")

Still, the USDA now reports that in 2014 (the latest numbers available as I write this) the average American consumed *only* 67 pounds of the sucrose and HFCS out of the 114 pounds the industry made available — slightly less than 60 percent. By doing so, a reasonably reliable number (114 pounds *delivered*) has been transformed into an unreliable number (67 pounds *consumed*). A number that can be used for historical trends and comparisons has been converted into a number that cannot.

The sugar industry prefers the latter, smaller number — "We perceive it to be in our interest to see as low a per-capita sweetener consumption estimate as possible," as one sugar industry executive wrote in a 2011 e-mail. The smaller number suggests that we don't eat or drink all that much sugar (or HFCS), after all. But it has no comparison. We have no meaningful way of adjusting sugar deliveries for loss decades or centuries ago. Nor can we use it to draw meaningful comparisons to the amount of other foods we supposedly consume today, because those *adjusted* numbers are also based on unreliable surveys and unsubstantiated assumptions.

For the sake of simplicity, I will typically refer in the text to the amount of sugar consumed per year (100 pounds per capita in the U.S. in 1920, for instance) because that's how it was referred to in the documents I cite, even though this number was technically the amount of sugar made available by industry, i.e., deliveries. When I refer to numbers that purport to be legitimate estimates of consumption, I will be explicit. It's a confusing business, but I'll do my best to keep it clear as we continue.

CHAPTER 1
DRUG OR FOOD?

The sweet shop in Llandaff in the year of 1923 was the very center of our lives. To us, it was what a bar is to a drunk, or a church is to a Bishop. Without it, there would have been little to live for. . . . Sweets were our life-blood.

ROALD DAHL,
Boy: Tales of Childhood, 1984

Imagine a moment when the sensation of honey or sugar on the tongue was an astonishment, a kind of intoxication. The closest I've ever come to recovering such a sense of sweetness was secondhand, though it left a powerful impression on me even so. I'm thinking of my son's first experience of sugar: the icing on the cake at his first birthday. I have only the testimony of Isaac's face to go by (that, and his fierceness to repeat the experience), but it was plain that his first encounter with sugar had intoxicated him — was in fact an ecstasy, in the literal sense of that word. That is, he was beside himself with the pleasure of it, no longer here with me in space and time in quite the same way he had been just a moment before. Between bites Isaac gazed up at me in amazement (he was on my lap, and I was delivering the ambrosial forkfuls to his gaping mouth) as if to exclaim, "Your world contains this? From this day forward I shall dedicate my life to it."

MICHAEL POLLAN,
Botany of Desire, 2001

What if Roald Dahl and Michael Pollan are right, that the taste of sugar on the tongue can be a kind of intoxication? Doesn't it suggest the possibility that sugar itself is an intoxicant, a drug? Imagine a drug that can intoxicate us, can infuse us with energy, and can do so when taken by mouth. It doesn't have to be injected, smoked, or snorted for us to experience its sublime and soothing effects. Imagine that it mixes well with virtually every food and particularly liquids, and that when given to infants it provokes a feeling of pleasure so profound and intense that its pursuit becomes a driving force throughout their lives.

Overconsumption of this drug may have long-term side effects, but there are none in the short term — no staggering or dizziness, no slurring of speech, no passing out or drifting away, no heart palpitations or respiratory distress. When it is given to children, its effects may be only more extreme variations on the apparently natural emotional roller coaster of childhood, from the initial intoxication to the tantrums and whining of what may or may not be withdrawal a few hours later. More than anything, our imaginary drug makes children happy, at least for the period during which they're consuming it. It calms their distress, eases their pain, focuses their attention, and then leaves them excited and full of joy until the dose wears off. The only downside is that children will come to expect another dose, perhaps to demand it, on a regular basis.

How long would it be before parents took to using our imaginary drug to calm their children when necessary, to alleviate pain, to prevent outbursts of unhappiness, or to distract attention? And once the drug became identified with pleasure, how long before it was used to celebrate birthdays, a soccer game, good grades at school? How long before it became a way to communicate love and celebrate happiness? How long before no gathering of family and friends was complete without it, before major holidays and celebrations were defined in part by the use of this drug to assure pleasure? How long would it be before the underprivileged of the world would happily spend what little money they had on this drug rather than on nutritious meals for their families?

How long would it be before this drug, as the anthropologist Sidney W. Mintz said about sugar, demonstrated "a near invulnerability to moral attack," before even writing a book such as this one was perceived as the nutritional equivalent of stealing Christmas?

What is it about the experience of consuming sugar and sweets, particularly during childhood, that invokes so readily the comparison to a drug? I have children, still relatively young, and I believe raising them would be a far easier job if sugar and sweets were not an option, if managing their sugar consumption did not seem to be a constant theme in our parental responsibilities. Even those who vigorously defend the place of sugar and sweets in modern diets — "an innocent moment of pleasure, a balm amid the stress of life," as the British journalist Tim Richardson has written — acknowledge that this does not include allowing children "to eat as many sweets as they want, at any time," and that "most parents will want to ration their children's sweets."

But why is it necessary? Children crave many things — Pokémon cards, *Star Wars* paraphernalia, *Dora the Explorer* backpacks — and many foods taste good to them. What is it about sweets that makes them so uniquely in need of rationing, which is another way of asking whether the comparison to drugs of abuse is a valid one?

This is of more than academic interest, because the response of entire populations to sugar has been effectively identical to that of children: once populations are exposed, they consume as much sugar as they can easily procure, although there may be natural limits set by culture and current attitudes about food. The primary barrier to more consumption — up to the point where populations become obese and diabetic and then, perhaps, beyond — has tended to be availability and price. (This includes, in one study, sugar-intolerant Canadian Inuit, who lacked the enzyme necessary to digest the fructose component of sugar and yet continued to consume sugary beverages and candy despite the "abdominal distress" it brought them.) As the price of a pound of sugar has dropped over the centuries — from the equivalent of 360 eggs in the thirteenth century to two in the early decades of the twentieth — the amount of sugar consumed has steadily, inexorably, climbed. In 1934, while sales of candy continued to increase during the Great Depression, *The New York Times* commented, "The depression proved that people wanted candy, and that as long as they had any money at all, they would buy it." During those brief periods of time during which sugar production surpassed our ability to consume it, the sugar industry and purveyors of sugar-rich products have worked diligently to increase demand and, at least until recently, have succeeded.

The critical question, what scientists debate, as the journalist and historian Charles C. Mann has elegantly put it, "is whether [sugar] is actually an addictive substance, or if people just act like it is." This question is not easy to answer. Certainly, people and populations have acted as though sugar is addictive, but science provides no definitive evidence. Until recently, nutritionists studying sugar did so from the natural perspective of viewing sugar as a nutrient — a carbohydrate — and nothing more. They occasionally argued about whether or not it might play a role in diabetes or heart disease, but not about whether it triggered a response in the brain or body that made us want to consume it in excess. That was not their area of interest.

The few neurologists and psychologists interested in probing the sweet-tooth phenomenon, or why we might need to ration our sugar consumption so as not to eat it to excess, did so typically from the perspective of how these sugars compared with other drugs of abuse, in which the mechanism of addiction is now relatively well understood. Lately, this comparison has received more attention as the public-health community has looked to ration our sugar consumption as a population, and has thus considered the possibility that one way to regulate these sugars — as with cigarettes — is to establish that they are, indeed, addictive. These sugars are very likely unique in that they are both a nutrient *and* a psychoactive substance with some addictive characteristics.

Historians have often considered the sugar-as-a-drug metaphor to be an apt one. "That sugars, particularly highly refined sucrose, produce peculiar physiological effects is well known," wrote the late Sidney

Mintz, whose 1985 book *Sweetness and Power* is one of two seminal English-language histories of sugar on which other, more recent writers on the subject (including myself) heavily rely.* But these effects are neither as visible nor as long-lasting as those of alcohol, or caffeinated beverages, "the first use of which can trigger rapid changes in respiration, heartbeat, skin color and so on." Mintz has argued that a primary reason that through the centuries sugar has escaped religious-based criticisms, of the kind pronounced on tea, coffee, rum, and even chocolate, is that, whatever conspicuous behavioral changes may occur when infants consume sugar, it did not cause the kind of "flushing, staggering, dizziness, euphoria, changes in the pitch of the voice, slurring of speech, visibly intensified physical activity, or any of the other cues associated with the ingestion" of these other drugs. As this book will argue, sugar appears to be a substance that causes pleasure with a price that is difficult to discern immediately and paid in full only years or

* The other is *The History of Sugar,* published in two encyclopedic volumes in 1949 and 1950, by Noël Deerr, a sugar-industry executive turned sugar historian.

decades later. With no visible, directly noticeable consequences, as Mintz says, questions of "long-term nutritive or medical consequences went unasked and unanswered." Most of us today will never know if we suffer even subtle withdrawal symptoms from sugar, because we'll never go long enough without sugar to find out.

Mintz and other sugar historians consider the drug comparison to be so fitting in part because sugar is one of a handful of "drug foods," to use Mintz's term, that came out of the tropics, and on which European empires were built from the sixteenth century onward, the others being, tea, coffee, chocolate, rum, and tobacco. Its history is intimately linked to that of these other drugs. Rum is distilled, of course, from sugarcane, whereas tea, coffee, and chocolate were not consumed with sweeteners in their regions of origin. In the seventeenth century, however, once sugar was added as a sweetener and prices allowed it, the consumption of these substances in Europe exploded. Sugar was used to sweeten liquors and wine in Europe as early as the fourteenth century; even cannabis preparations in India and opium-based wines and syrups included sugar as a major ingredient.

Kola nuts, containing both caffeine and

traces of a milder stimulant called theobromine, became a product of universal consumption in the late nineteenth century, first as a coca-infused wine in France (Vin Mariani) and then as the original mixture of cocaine and caffeine of Coca-Cola, with sugar added to mask the bitterness of the other two substances. The removal of the cocaine in the first years of the twentieth century seemed to have little influence on Coca-Cola's ability to become, as one journalist described it in 1938, the "sublimated essence of all that America stands for," the single most widely distributed product on the planet and the second-most-recognizable word on Earth, "okay" being the first. It's not a coincidence that John Pemberton, the inventor of Coca-Cola, had a morphine addiction that he'd acquired after being wounded in the Civil War. Coca-Cola was one of several patent medicines he invented to help wean him off the harder drug. "Like Coca, Kola enables its partakers to undergo long fast and fatigue," read one article in 1884. "Two drugs, so closely related in their physiological properties, cannot fail to command early universal attention."

As for tobacco, sugar was, and still is, a critical ingredient in the American blended-tobacco cigarette, the first of which was Camel, introduced by R. J. Reynolds in 1913. It's this "marriage of tobacco and sugar," as a sugar-industry report described it in 1950, that makes for the "mild" experience of smoking cigarettes as compared with cigars and, perhaps more important, makes it possible for most of us to inhale cigarette smoke and draw it deep into our lungs. It's the "inhalability" of American blended cigarettes that made them so powerfully addictive — as well as so potently carcinogenic — and that drove the explosion in cigarette smoking in the United States and Europe in the first half of the twentieth century, and the rest of the world shortly thereafter, and, of course, the lung-cancer epidemics that have accompanied it.

Unlike alcohol, which was the only commonly available psychoactive substance in the Old World until sugar, nicotine, and caffeine arrived on the scene, the latter three had at least some stimulating properties, and so offered a very different experience, one that was more conducive to the labor of everyday life. These were the "eighteenth-century equivalent of uppers," writes the Scottish historian Niall Ferguson. "Taken together, the new drugs gave English society an almighty hit; the Empire, it might be said, was built on a huge sugar, caffeine and nicotine rush — a rush nearly everyone could experience."

Sugar, more than anything, seems to have made life worth living (as it still does) for so many, particularly those whose lives were absent the kind of pleasures that relative wealth and daily hours of leisure might otherwise provide. As early as the twelfth century, one contemporary chronicler of the Crusades, Albert of Aachen, was describing merely the opportunity to sample the sugar from the cane that the Crusaders found growing in the fields of what are now Israel and Lebanon as in and of itself "some compensation for the sufferings they had endured." "The pilgrims," he wrote, "could not get enough of its sweetness."

As sugar, tea, and coffee instigated the transformation of daily life in Europe and the Americas in the seventeenth and eighteenth centuries, they became the indulgences that the laboring classes could afford; by the 1870s, they had come to be considered necessities of life. During periods of economic hardship, as the British physician and researcher Edward Smith observed at the time, the British poor would sacrifice the nutritious items of their diet before they'd cut back on the sugar they consumed. "In nutritional terms," suggested three British researchers in 1970 in an analysis of the results of Smith's survey, "it would have been better if some of the money spent on sugar had been diverted to buy bread and potatoes, since this would have given them very many more calories for the same money, as well as providing some protein, vitamins and minerals, which sugar lacks entirely. In fact however we find that a taste for the sweetness of sugar tends to become fixed. The choice to eat almost as much sugar as they used to do, while substantially reducing the amount of meat, reinforces our belief that people develop a liking for sugar that becomes difficult to resist or overcome."

Sugar was "an ideal substance," says Mintz. "It served to make a busy life seem less so; in the pause that refreshes, it eased, or seemed to ease the changes back and forth from work to rest; it provided swifter sensations of fullness or satisfaction than complex carbohydrates did; it combined easily with many other foods, in some of which it was also used (tea and biscuit, coffee and bun, chocolate and jam-smeared bread). . . . No wonder the rich and powerful liked it so much, and no wonder the poor learned to love it." What Oscar Wilde wrote about a cigarette in 1891, when that indulgence was about to explode in popularity and availability, might also be said about sugar: It is "the perfect pleasure. It is exquisite, and it leaves one unsatisfied. What more can one want?"

Sugar craving does seem to be hard-wired in our brains. Children certainly respond to it instantaneously, from birth (if not in the womb) onward. Give babies a choice of sugar water or plain, wrote the British physician Frederick Slare three hundred years ago, and "they will greedily suck down the one, and make Faces at the other: Nor will they be pleas'd with Cows Milk, unless that be bless'd with a little Sugar, to bring it up to the Sweetness of Breast-Milk." Slare's observation was confirmed experimentally in the early 1970s by Jacob Steiner, a professor of oral biology at the Hebrew University of Jerusalem. Steiner studied and photographed the expressions of newborn infants given a taste of sugar water even before they had received breast milk or any other nourishment. The result, he wrote, was "a marked relaxation of the face, resembling an expression of 'satisfaction,' often accompanied 'by a slight smile,' " which was almost always followed "by an eager licking of the upper lip, and sucking movements."

When Steiner repeated the experiment with a bitter solution, the newborns spit it out.

This raises the question of why humans evolved a sweet tooth, requiring intricate receptors on the tongue and the roof of the mouth, and down into the esophagus, that will detect the presence of even minute amounts of sugar and then signal this taste via nerves extending up into the brain's limbic system. Nutritionists usually answer by saying that in nature a sweet taste signaled either calorically rich fruits or mother's milk (because of the lactose, a relatively sweet carbohydrate, which can constitute up to 40 percent of the calories in breast milk), so that a highly sensitive system for distinguishing such foods and differentiating them from the tastes of poisons, which we recognize as bitter, would be a distinct evolutionary advantage. But if caloric or nutrient density is the answer, the nutritionists and evolutionary biologists have to explain why fats do not also taste sweet to us. They have twice as many calories per gram as sugars do (and more than half the calories in mother's milk come from fat).

One proposition commonly invoked to explain why the English would become the world's greatest sugar consumers and remain so through the early twentieth century, alongside the fact that the English had the world's most productive network of sugar-producing colonies, is that they had lacked any succulent native fruit, and so had little previous opportunity to accustom themselves to sweets, as Mediterranean populations did. As such, the sweet taste was more of a novelty to the English, and their first exposure to sugar, as this thinking goes, occasioned more of a population-wide astonishment. According to this argument, Americans then followed the British so closely as sugar consumers because the original thirteen colonies were settled by the English, who brought their sweet cravings with them. The same explanation holds for Australians, who had caught up to the British as sugar consumers by the early decades of the twentieth century.

All of this is speculation, however, as is the notion that it was the psychoactive aspects of sugar consumption that provided the evolutionary advantage. The taste of sugar will soothe distress, and thus "distress vocalizations" in infants; consuming sugar will allow adults to work through pain and exhaustion and to assuage hunger pains. That sugar works as a painkiller or at least a powerful distraction to infants is evidenced by its use during circumcision ceremonies

— even in hospitals on the day after birth — to soothe and quiet the newborn. If sugar, though, is only a distraction to the infant and not actively a pain reliever or a psychoactive inducer of pleasure that overcomes any pain, as this view posits, we have to explain why in clinical trials it is more effective in soothing the distress of infants than the mother's breast and breast milk itself.

Many animals do respond positively to sugar — they have a sweet tooth — but not all. Cats don't, for instance, but they're obligate carnivores (in nature, they eat only other animals). Chickens don't, nor do armadillos, whales, sea lions, some fish, and cowbirds. Despite the ubiquitous use of rats in the research on sugar addiction, some strains of laboratory rats prefer maltose — the carbohydrate in beer — to sugar. Cattle, on the other hand, will happily fatten themselves on sugar, an observation that was made in the late nineteenth century, when the price of sugar fell sufficiently that farmers could afford to use it for feed. In one study published in 1952, agronomists reported that they could get cattle to eat plants they otherwise disdained by spraying the plants with sugar or molasses (the cattle preferred the latter) — in other words, by sugar-coating them. "In several instances," the researchers reported, "the cattle quickly became aware of what was going on and followed the spraying can around expectantly." The cattle had the same response to artificial sweeteners, suggesting that "the cattle liked anything sweet whether it had food value or not." By sweetening with sugar, as an essay in *The New York Times* observed in 1884, "we can give a false palatableness to even the most indigestible rubbish."

The actual research literature on the question of whether sugar is addictive and thus a nutritional variation on a drug of abuse is surprisingly sparse. Until the 1970s and for the most part since then, mainstream authorities have not considered this question to be particularly relevant to human health. The very limited research allows us to describe what happens when rats and monkeys consume sugar, but we're not them and they're not us. The critical experiments are rarely if ever done in humans, and certainly not children, for the obvious ethical reasons: we can't compare how they respond to sugar, cocaine, and heroin, for instance, to determine which is more addictive.

Sugar does induce the same responses in

the region of the brain known as the "reward center" — technically, the nucleus accumbens — as do nicotine, cocaine, heroin, and alcohol. Addiction researchers have come to believe that behaviors required for the survival of a species — specifically, eating and sex — are experienced as pleasurable in this part of the brain, and so we do them again and again. Sugar stimulates the release of the same neurotransmitters — dopamine in particular — through which the potent effects of these other drugs are mediated. Because the drugs work this way, humans have learned how to refine their essence into concentrated forms that heighten the rush. Coca leaves, for instance, are mildly stimulating when chewed, but powerfully addictive when refined into cocaine; even more so taken directly into the lungs when smoked as crack cocaine. Sugar, too, has been refined from its original form to heighten its rush and concentrate its effects, albeit as a nutrient that provides energy as well as a chemical that stimulates pleasure in the brain.

The more we use these substances, the less dopamine we produce naturally in the brain, and the more habituated our brain cells become to the dopamine that *is* produced — the number of "dopamine receptors" declines. The result is a phenomenon known as dopamine down-regulation: we need more of the drug to get the same pleasurable response, while natural pleasures, such as sex and eating, please us less and less. The question, though, is what differentiates a substance that works in the reward center to trigger an intense experience of pleasure and yet isn't addictive, and one that happens to be both. Does sugar cross that line? We can love sex, for instance, and find it intensely pleasurable without being sex addicts. Buying a new pair of shoes, for many of us, will also stimulate a dopamine response in the reward center of the brain and yet not be addictive.

Rats given sweetened water in experiments find it significantly more pleasurable than cocaine, even when they're addicted to the latter, and more than heroin as well (although the rats find this choice more difficult to make). Addict a rat over the course of months to intravenous boluses of cocaine, as the French researcher Serge Ahmed has reported, and then offer it the choice of a sweet solution or its daily cocaine fix, and the rat will switch over to the sweets within two days. The choice of sweet taste over cocaine, Ahmed reports, may come about because neurons in the brain's reward circuitry that respond specifically to sweet taste outnumber those that respond to cocaine fourteen to one; this general finding has been replicated in monkeys.

This animal research validates the anecdotal experience of drug addicts and alcoholics, and the observations of those who both study and treat addiction, that sweets and sugary beverages are valuable tools — "sober pleasures" — to wean addicts off the harder stuff, perhaps transferring from one addiction, or one dopamine-stimulating substance, to another, albeit a relatively more benign one. "There is little doubt that sugar can allay the physical craving for alcohol," as the neurologist James Leonard Corning observed over a century ago. The twelve-step bible of Alcoholics Anonymous — called the Big Book — recommends the consumption of candy and sweets in lieu of alcohol when the cravings for alcohol arise. Indeed, the per capita consumption of candy in the United States doubled with the beginning of Prohibition in 1919, as Americans apparently turned en masse from alcohol to sweets. Ice-cream consumption showed a "tremendous increase" coincident with Prohibition. By 1920, sugar consumption in the United States hit record highs, while breweries were being converted into candy factories. "The wreckage of the liquor business," *The New York Times* reported, "is being salvaged for the production of candy, ice cream and syrups." Five years later, British authorities suggested that this tremendous increase in ice-cream consumption "due to prohibition was injurious to health," but an American college president countered that the trade-off was apparently worth it, as he had "never heard of a man who ate excessive quantities of the confection going home to beat his wife."

All of this is worth keeping in mind when we think about how inexorably sugar and sweets came to saturate our diets and dominate our lives, as the annual global production of sugar increased exponentially from the 1600s onward. The yearly amount of sugar consumed per capita more than quadrupled in England in the eighteenth century, from four pounds to eighteen, and then more than quadrupled again in the nineteenth. In the United States, yearly sugar consumption increased sixteen-fold over that same century.

By the early twentieth century, sugar had assimilated itself into all aspects of our eating experience — consumed during breakfast, lunch, dinner, and snacks. Nutritional authorities were already suggesting what appeared to be obvious, that this increased consumption was a product of at least a kind of addiction — "the development of the sugar appetite, which, like any other appetite — for instance, the liquor appetite — grows by gratification."

A century later still, sugar has become an ingredient avoidable in prepared and packaged foods only by concerted and determined effort, effectively ubiquitous: not just in the obvious sweet foods — candy bars, cookies, ice creams, chocolates, sodas, juices, sports and energy drinks, sweetened iced tea, jams, jellies, and breakfast cereals (both cold and hot) — but also in peanut butter, salad dressing, ketchup, barbecue sauces, canned soups, cold cuts, luncheon meats, bacon, hot dogs, pretzels, chips, roasted peanuts, spaghetti sauces, canned tomatoes, and breads. From the 1980s onward, manufacturers of products advertised as uniquely healthy because they were low in fat or specifically in saturated fat (not to mention "gluten free, no MSG & 0g trans fat per serving") took to replacing those fat calories with sugar to make them equally, if not more, palatable, and often disguising the sugar under one or more of the fifty-plus names by which the fructose-glucose combination of sugar and high-fructose corn syrup might be found. Fat was removed from candy bars, sugar added or at least kept, so that they became health-food bars. Fat was removed from yogurts and sugars added, and these became heart-healthy snacks, breakfasts, and lunches. It was as though the food industry had decided en masse, or its numerous focus groups had sent the message, that if a product wasn't sweetened at least a little, our modern palates would reject it as inadequate and we would purchase instead a competitor's version that was.

Along the way, sugar and sweets became synonymous with love and affection and the language with which we communicate them — "sweets," "sweetie," "sweetheart," "sweetie pie," "honey," "honeybun," "sugar," and all manner of combinations and variations. Sugar and sweets became a primary contribution to our celebrations of holidays and accomplishments, both major and minor. For those of us who don't reward our existence with a drink (and for many of us who do), it's a candy bar, a dessert, an ice-cream cone, or a Coke (or Pepsi) that makes our day. For those of us who are parents, sugar and sweets have become the tools we wield to reward our children's accomplishments, to demonstrate our love and our pride in them, to motivate them, to entice them. Sweets have become the currency of childhood and of parenting.

The common tendency is, again, to think of this transformation as driven by the mere fact that sugars and sweets taste good. We can call it the "pause that refreshes" hypothesis of sugar history. The alternative way to think about this is that sugar took over our diets because the first taste, whether for an infant today or for an adult centuries ago, is literally, as Michael Pollan put it, an astonishment, a kind of intoxication; it's the kindling of a lifelong craving, not identical but analogous to that of other drugs of abuse. Because it is a nutrient, and because the conspicuous sequelae of its consumption are relatively benign compared with those of nicotine, caffeine, and alcohol — at least in the short term and in small doses — it remained, as Sidney Mintz says, nearly invulnerable to moral, ethical, or religious attacks. It remained invulnerable to health attacks as well.

Nutritionists have found it in themselves to blame our chronic ills on virtually any element of the diet or environment — on fats and cholesterol, on protein and meat, on gluten and glycoproteins, growth hormones and estrogens and antibiotics, on the absence of fiber, vitamins, and minerals, and surely on the presence of salt, on processed foods in general, on overconsumption and sedentary behavior — before they'll concede that it's even possible that sugar has played a unique role in any way other than merely getting us all to eat (as Harvard's Fred Stare put it forty years ago) too damn much. And so, when a few informed authorities over the years did, indeed, risk their credibility by suggesting sugar was to blame, their words had little effect on the beliefs of their colleagues or on the eating habits of a population that had come to rely on sugar and sweets as the rewards for the sufferings of daily life.

Chapter 2
The First Ten Thousand Years

M. Delacroix, a writer as charming as he is prolific, complained once to me at Versailles about the price of sugar, which at that time cost more than five francs a pound. "Ah," he said in a wistful, tender voice, "if it can ever again be bought for thirty cents, I'll never more touch water unless it's sweetened!" His wish was granted.

JEAN ANTHELME BRILLAT-SAVARIN
The Physiology of Taste, 1825

Sugar is a fuel for plants and can be found in all of them — in some, however, more than in others. It's a safe bet that humans have tried to extract sugar, at one time or another, from pretty much every substance or plant that was noticeably sweet and held the promise of offering its sugar up in quantity. Honey was consumed throughout Europe and Asia before sugar displaced it, and when European colonists arrived in the New World and found no honey, they introduced honeybees, which Native Americans took to calling the "English Man's Fly." Native Americans were using maple syrup as a sweetener before the Europeans arrived, and they introduced the colonists to the taste. (Thomas Jefferson was a proponent of maple syrup because it rendered slave labor unnecessary. The sugar maple, he wrote, "yields a sugar equal to the best from the cane, yields it in great quantity, with no other labor than what the women and girls can bestow. . . . What a blessing.") But neither maple syrup nor honey can be used to sweeten cold beverages, and neither mixes well with coffee. Neither could be produced in the quantities necessary to compete with sugar. We still consume them, but in limited quantities and for limited uses.

Even sorghum, an Old World grass used as cattle feed in Africa and chewed by villagers there for its sweetness, had a run in the late nineteenth century as a potential source of sugar, a competitor to cane and beet sugar. The U.S. Department of Agriculture took it up and "kindled an enthusiasm that amounted to a craze," but droughts and insect visitations did it in. Cane and then beet sugar and now high-fructose corn syrup simply won out, in that they were the sweeteners that could be mass-produced economically and provided in quantities necessary to satisfy what appears to have been an almost limitless demand.

Anthropologists believe that sugarcane itself was first domesticated in New Guinea about ten thousand years ago. As evidence that it was revered even then, creation myths in New Guinea have the human race emerging from the sexual congress of the first man and a stalk of sugarcane. The plant is technically a grass, growing to heights of twelve to fifteen feet, with juicy stalks that can be six inches around. In tropical soils, sugarcane will grow from cuttings of the stem, and will ripen or mature in a year to a year and a half. The juice or sap from the cane, at least the modern variety, is mostly water and as much as 17 percent sugar. This makes the cane sweet to chew but not intensely so. Anthropologists assume that early farmers domesticated the cane for the sweetness to be derived from chewing the stalks *and* the energy it provided. Well before the art of refining came along, sugarcane domestication had already spread to India, China, the Philippines, and Indonesia.

Without refining, the juice of sugarcane is for local consumption only. Within a day of cutting, the sugarcane stalks will begin to ferment and then rot. But the juice can be squeezed or crushed or pounded out of the cane, and that, in turn, as farmers in northern India discovered by around 500 B.C., can be transformed into a raw sugar by cycles of heating and cooling — a "series of liquid-solid operations." The sugar crystallizes as the liquid evaporates. One end product is molasses, a thick brown viscous liquid; another, requiring greater expenditures of time and effort, is dry crystalline sugar of colors ranging from brown to white. The greater the refining effort, the whiter and more pure is the end product.

When cultivated with the instruments of modern technology, sugarcane can produce (as the sugar industry and nutritionists would state in its defense repeatedly in the twentieth century) more calories per acre to feed a population than any other animal or plant. It can survive years of storage; it travels well; it can be consumed on arrival unheated and uncooked. And, unlike honey or maple syrup, it has no distinctive taste or aftertaste. Refined sugar is colorless and odorless. It is nothing more than the crystallized essence of sweet. Other than salt, it is the only pure chemical substance that humans consume. And it provides four calories of energy per gram.

Sugar is extraordinarily useful in food preparation, even when sweetness is not necessarily the desired result, and this is one reason why sugar in all its various names and forms is now ubiquitous in modern processed foods. Sugar allows for the preservation of fruits and berries by inhibiting the growth of micro-organisms that would otherwise cause spoiling. As such, inexpensive sugar made possible the revolution in jams and jellies that began in the mid-nineteenth century (one of many revolutions in sugar-rich foods that began at the same time, as we'll discuss shortly). It inhibits mold and bacteria in condensed milk and other liquids by increasing what's called the osmotic pressure of the liquid. It reduces the harshness of the salt that's used for curing and preserving meat (and the salt increases the sweetness of the sugar). Sugar is an ideal fuel for yeast, and thus the rising and leavening of bread. The caramelization of sugar provides the light-brown colors in the crust of bread. Dissolve sugar in water and it adds not only sweetness but viscosity, and thus creates the body and what food scientists call the "mouth feel" of a soda or juice. As a seasoning or a spice, it enhances flavors already present in the food, decreases bitterness, and improves texture.

All of this was assuredly secondary to sweetness and nourishment, and perhaps any perceived medicinal use, when sugar began its dispersion throughout the world two thousand years ago. From India, Buddhist missionaries carried it to China and Japan. Muslim explorers then discovered sugar in China and carried it back to Arabia via Persia shortly before the Muslim expansion that began in the seventh century after the death of Muhammad. As the story goes, Chosroes I, Emperor of Persia, asked for a drink of water from a young girl in a garden, and she gave him a cup of sugarcane juice chilled with snow. Chosroes promptly asked for a refill and then contemplated stealing the garden while she was gone. "I must remove these people elsewhere and take this garden for myself," he said to himself. Whether he did or not, Chosroes is credited with taking the sugarcane back to Persia, and the Muslim Empire then spread sugarcane-growing around the Mediterranean — to Malta, Sicily, Cyprus, southern Spain, and North and East Africa.

By the tenth century, the two great sugar-producing areas outside of India and China were at the head of the Persian Gulf in the Tigris-Euphrates delta, and in the Nile River Valley in Egypt. It was the Egyptians who first developed the refining techniques that have been used more or less ever since. Records exist of the use of sugar at that time in the royal households of Egyptian viziers and caliphs to the tune of a thousand pounds per *day,* and of Ramadan feasts in which seventy-five tons of sugar were used at a single celebration, much of it to sculpt table decorations that were either consumed outright or given to the neighborhood beggars after the feasts.

Sugar began to seep into Northern Europe with the Crusades in the eleventh century. When the first Crusaders made it back home, they told stories about the fields of sugarcane they had seen and the locals, as Albert of Aachen recorded, "sucking enthusiastically on these reeds, delighting themselves with their beneficial juices, and seem[ing] unable to sate themselves with the pleasure." By then the Crusaders were overseeing sugar production in the areas they had conquered. Sugar was "a most precious product, very necessary for the use and health of mankind," wrote one contemporary chronicler. When Crusaders with a taste for sugar returned home, Italian city-states began shipping sugar by land and sea routes to Northern Europe and the British Isles. Sugar appears in the kitchen expenditures of Henry II at the tail end of the twelfth century, listed as a spice; this was among the first mentions ever of sugar use in Britain. In 1288, Edward I's household used over sixty-two hundred pounds of sugar.

As sugar diffused through Europe, it did so primarily as a medicine — as would tea, coffee, tobacco, and chocolate centuries later — a decorative, a spice, and a preservative. (Edward I's delicate son, who suffered perpetually from colds, was given sugar and sugar sticks as part of his treatment — "to no avail, as he died early.") In the thirteenth century, Thomas Aquinas said sugar consumption did not have to be prohibited during fasts because sugar was not "eaten with the end in mind of nourishment, but rather for ease in digestion; accordingly, they do not break the fast any more than taking of any other medicine." For the next five hundred years, sugar would be ingested medicinally as much as for any other use. "It was good for almost every part of the body, for the very young, for the very old, for the sick and for the healthy," wrote the British historian James Walvin. "It cured and prevented illnesses; it refreshed the weary, invigorated the weak."

As the price of sugar slowly dropped, its use as a sweetener and a food went up. It moved from the shops of apothecaries, "who kept it exclusively for invalids," to being devoured "out of gluttony." By the fourteenth century, sugar was appearing in cooking recipes; by the fifteenth, it was an indispensable ingredient in the kitchens of those wealthy enough to afford it. "No food refuses, so to speak, sugar," is how one Italian gastronome described it at the time, an opinion that is supported by the existence of several recipes from medieval English cuisine for sugar-sprinkled oysters. "Sugar spoils no dish," was a mid-sixteenth-century German variation on the same notion.

The barriers to the increased consumption of sugar, as I suggested earlier, would invariably be cost and availability, which in turn were constrained by land and labor. Sugarcane itself can be grown only in or near the tropics; it needs warm weather, and either a lengthy rainy season or extensive irrigation to provide the considerable water necessary. Wherever sugar could be grown in the Old World it was grown, but the land was limited; planting, harvesting, and refining sugar, and in sufficient quantities to sell anywhere other than at local markets, was not work that could be done by individual peasant farmers. It required mills for extracting the juice from the cane; vessels and copious wood for boiling; pots for crystallizing; containers for shipping and storing; and facilities for transport.

The work itself was dreadful, as Charles C. Mann has described it — "swinging machetes into the hard, soot-smeared cane under the tropical sun, [splattering the field hands] head to foot with a sticky mixture of dust, ash, and cane juice," not to mention working the mills and the infernolike refineries or "sugar factories," as they were then called. It was difficult to find a population poor enough and desperate enough to do it willingly.

Slaves, having no choice in the matter, became the solution. If nothing else, the intimate relationship between slavery and sugar would demonstrate what atrocities our ancestors were willing to tolerate and perpetrate for the sake of their sweet tooth, their sugar rushes, and the money to be made by satisfying them.

Sugar and slavery went hand in hand from the earliest times. When Muslims began growing sugar in the Middle East in the seventh century, they imported black slaves from East Africa to work the fields. Slaves were apparently used throughout the Mediterranean sugar industry, often working beside peasant labor. As Portugal and then Spain sent ships progressively south along the African coast in the early fifteenth century, inaugurating the Age of Discovery, they simultaneously began trading in black slaves and putting them to work in the sugar plantations on the newly colonized islands in the nearby Atlantic — Madeira, the Azores, the Cape Verde Islands, São Tomé, Principe and Annobon, and the Canary Islands.

It was Columbus who first brought sugar to the New World — on his second voyage, in 1493, having stopped first in the Canary Islands, where he picked up both sugarcane plantings and "field experts in cultivation" who could grow the sugar. The sugarcane grew with Biblical speed in the fertile soil of Hispaniola (now Haiti and the Dominican Republic) — sprouting in seven days, Columbus reported — but the planters themselves sickened and died, as did the Amerindian slaves used for labor. In 1506, Canary Island sugarcane was brought back to Hispaniola, and every inhabitant who would "erect a sugar mill should have five hundred pieces of eight in gold lent him." Ten years later, loaves of sugar were being sent back to Spain as gifts to the emperor; by 1525, the trade was "so lucrative that sugar was shipped along with treasure and pearls under convoy."

Columbus's pilot, Pinzón, brought sugar to Brazil with his voyage of discovery in 1499, and the Portuguese colonists in Brazil created the first viable sugar industry in the New World. By 1526, sugar was being refined in a factory and sent back to Portugal, making sugar the first agricultural commodity to be shipped in commercial quantities from the New World to the Old. Brazilian sugar dominated the trade in the sixteenth century. Sugar factories sprang up throughout the country. By the end of the century, they were exporting back to Europe at least ten thousand pounds of sugar each year — by some estimates, tens of thousands of pounds.

In Mexico, the first Spanish conquistadors, in the early sixteenth century, brought sugar with them as well. They founded a nascent sugar industry as they marched through the region. Cortés himself gets credit not only for conquering the Aztec Empire (with the considerable help of smallpox and other infectious diseases), but also for erecting two of the earliest sugar mills on the continent. By 1552, when Gonzalo Fernández de Oviedo published his *History of the Conquest of Mexico,* he insisted that the fledgling Mexican sugar industry was capable of producing enough sugar "to supply the whole of Christendom." The conquistadors also came upon the natives drinking chocolate, although unsweetened and spiced with chili peppers. The Spaniards found the drink unpleasant — "better to be tossed out to pigs than drunk by men" — but Cortés sent a gift of cocoa beans back to Emperor Charles V in 1527 nonetheless. By the end of the century, Spanish aristocrats were mixing their chocolate with sugar and drinking sweetened hot chocolate morning and afternoon.

Both the Spaniards and Portuguese first used the natives of the Americas to work their sugar plantations, but the forced labor and epidemic diseases brought over from Europe and Africa decimated these populations. And so they shipped in African slaves to work the plantations in the New World. When the French and British established colonies in the Caribbean in the seventeenth century, they, too, entered the sugar business, depending on slave labor from Africa to do the backbreaking labor of harvesting sugarcane on their plantations.

The British had tried to grow sugarcane on their first permanent colony in the New World at Jamestown, Virginia, in 1607, but the climate wasn't suitable. The British succeeded in Barbados in the 1640s and later Jamaica, only after Dutch refugees from Brazil — sugar-industry veterans — brought the sugarcane with them and taught the British how to grow and refine it.* The number of slaves on Barbados, the richest of the sugar islands until Jamaica later eclipsed it, went from a handful early in the seventeenth century to more than forty-six thousand in 1683. By the 1830s, when the British emancipationists finally put an end to the slave trade, some twelve and a half million Africans had been shipped off as slaves to the New World; two-thirds of them worked and died growing and refining sugar.

From the seventeenth through the nineteenth centuries, sugar was the equivalent, economically and politically, of oil in the

* The Dutch had initially conquered northern Brazil, after a decade-long struggle that concluded in 1635, motivated by the profits to be made growing sugar there. The Portuguese tossed them out in 1654, and it was these Dutch refugees who settled in Barbados and Jamaica.

twentieth. It was the stuff over which wars were fought, empires built, and fortunes made and lost. By 1775, "King Sugar," or "white gold," as it was known, constituted almost a fifth of all British imports, five times that of tobacco. The result, as the historian of science Robert Proctor has written about tobacco and taxation, was a "second addiction" — both the British and U.S. governments came to be vigorous promoters of the sugar industry because of the revenues they could garner by taxing it. Sugar was an ideal target of taxation: production was localized to tropical colonies, so its import could be controlled, and it was in universal demand but not (yet) considered a necessity of life. (The same was true of tea; the sweetening of tea and the burgeoning tea industry in India also drove sugar consumption through the British Empire in this era.) The British government began taxing sugar imports from the Caribbean, along with tobacco, in the late seventeenth century. The Americans followed a century later, after the Revolution, and after realizing how much money could be raised from sugar to help get a fledgling country on its feet.

For the sugar islands in the Caribbean, sugar production was so profitable that it seemed worthwhile to grow almost exclusively sugar and to import anything else needed for life. American colonies then thrived on the business of providing the necessities, the basic foodstuffs, which these sugar colonies failed to produce. Indeed, a primary reason the British West India Company had set out in the 1660s to wrest New York City (then New Amsterdam) from the Dutch was that it needed a port on the American mainland — an entrepôt — "from which they could obtain slaves and food in exchange for raw sugar and molasses." When the Dutch agreed to let the British keep New York in 1667, it was in exchange for Dutch Guiana (now Suriname) and its then more valuable sugar plantations. Not until the 1790s were Americans successfully growing any sugarcane — in Louisiana — although already sugar refineries, turning raw sugar from the Caribbean into refined sugar, were proliferating up and down the Northeastern coast. By 1810, thirty-three refineries were operating; by 1860, eighteen were operating in New York alone.

Many of the wealthiest New York families would make their fortunes initially as sugar refiners, as confectioners, and as middlemen in the triangular slave trade that hauled sugar and molasses north to New York, sent rum to Africa, and brought slaves back to the Caribbean, while also supplying the sugar islands in the Caribbean directly with the food and naval stores "without which the West Indian plantations couldn't survive." And it was the British decision in 1764 to enforce a tax on molasses in the colonies that helped incite the revolutionary feelings that would lead to independence. "I know not why we should blush to confess that molasses was an essential ingredient in American independence," wrote John Adams in 1775. "Many great events have proceeded from much smaller causes."

Sidney Mintz has elegantly described the arc of sugar's early history as that of a "luxury of kings into the kingly luxury of commoners." That transformation had been completed in the United Kingdom by the early nineteenth century, when sugar consumption per capita was approaching twenty pounds per year. The decades that followed would transform sugar into as much an article of necessity in life as bread itself. The latter stage in this transformation was marked in England in 1874, when the government finally abolished import duties, on the basis that sugar had become, as one member of Parliament described it, "the delight of childhood and the solace of old age," besides being "exceedingly nutritious and wholesome"; so, by this logic, the poor should have every right to consume as much as did the rich. In 1890 when the U.S. Congress was debating the same question — whether to repeal the tax on imported sugar, which it would never do — *The New York Times* noted that more than half a billion dollars had been collected in sugar taxes by the federal government in the 1880s alone.

Two factors ultimately drove this final transformation of sugar from a luxury for the wealthy to a pleasure for all. One was the development of the beet-sugar industry, representing a source of sugar that could be grown outside the tropics, in temperate climates. In the United States, this meant a two-thousand-mile-wide, north-to-south swath that stretched from coast to coast. In Europe and Asia, it meant a domestic supply of sugar for all those countries — including, most notably, Germany, Austria, and Russia — that had no access to the tropics or tropical colonies.

German chemists had succeeded in extracting and refining sugar from selected white beets as early as the 1740s, but they failed to make it profitable. ("To scientific ability he did not unite business acumen," wrote Noël Deerr in *The History of Sugar* about the first of these German beet-sugar entrepreneurs.) In 1811, when the British blockade of Europe during the Napoleonic Wars cut off the sugar supply to France, a French naturalist and banker named Benjamin Delessert succeeded at both refining sugar from beets and doing so in a way that wouldn't lead to bankruptcy. Napoleon famously traveled to Delessert's sugar factory to give him the medal of the Legion of Honor. In a speech to the French chambers of commerce, Napoleon suggested that the English could now throw their cane sugar "into the Thames," because they wouldn't be selling it on the Continent anymore. Napoleon allotted eighty thousand acres for growing sugar beets and established technical centers to teach the art and business of beet-sugar production. Within three years, over three hundred factories were producing beet sugar in France alone.

Napoleon's beet-sugar revolution would be temporarily derailed with his defeat in 1814 and the end of the continental blockade by the British. Once cheap sugar from the Caribbean flowed back into Europe, beet-sugar manufacturers couldn't compete with the lower prices. However, the abolition of slavery by the English in the 1830s, and the temporary collapse of the Caribbean sugarcane industry that followed, gave European beet-sugar producers another opportunity to get the industry up and running. By the late 1850s, sugar from beets coming out of Europe and Russia constituted more than 15 percent of world sugar production. By 1880, beet sugar had surpassed cane sugar, and the total amount of all sugar being refined and apparently consumed worldwide had increased over fivefold in forty years.

When the U.S. Department of Agriculture was founded in 1862, its impetus, as much as anything, was to encourage sugar-beet production.* Among its first acts was to analyze different strains of beets for their sugar content. Six years later, the commissioner of agriculture was claiming that it was only because of the U.S. government's

* The influence of science in the sugar industry cannot be underestimated. According to Deborah Jean Warner, a curator at the National Museum of American History and author of *Sweet Stuff,* beet sugar was the first agricultural endeavor to rely on scientific expertise to generate higher yields and strive for quality control, and when the American Chemical Society was founded in 1876, most of the founding members were sugar chemists.

encouragement of the fledgling beet-sugar industry that it might now "be numbered among the industries which bless the world."

The second factor in the transformation of sugar into a dietary staple — one of life's necessities — was technology. The industrial revolution, inaugurated by Watt's steam engine in 1765, transformed sugar production and refining just as it did virtually every other existing industry in the nineteenth century. By the 1920s, sugar refineries were producing as much sugar in a single *day* — millions of pounds — as would have taken refineries in the 1820s an entire decade.

With sugar becoming so cheap that everyone could afford it, the manner in which we consumed it would change as well. Not only did we add sugar to hot beverages and bake it into wheat products or spread it on top — jams and jellies were two foods that cheap, available sugar made ubiquitous, since fruit could now be preserved at the end of the growing season and provide nutrition (sweetened, of course) all year round — but the concept of a dessert course emerged for the first time in history in the mid-nineteenth century, the expectation of a serving of sweets to finish off a lunch or dinner. The industrial work break also emerged, as a new era of factory workers learned to partake of some combination of nicotine, caffeine, and sugar; cigarettes, coffee and tea, and sweetened biscuits or candy could all be purchased inexpensively.

The food entrepreneurs of the era, taking advantage of the industrial tools now available, created entirely new foods that could be mass-produced and sold everywhere in unprecedented quantities. In the 1840s, as Mark Twain wrote of his youth in rural Missouri, both sugar and molasses were bought in bulk out of barrels at the village store. Conspicuously absent from Twain's vivid enumeration of the items for sale in his uncle's country store in his hometown of Florida, Missouri, were *any* of the mass-produced foods or drinks through which we consume sugar today: no candy, ice cream, chocolate bars, packaged cakes or cookies, sodas, or juices. All of those would be effectively invented in the next half-century, as would the industries that would mass-produce them, the railroads that would ship them nationwide, the bottling and packaging needed to contain them, the labels to go on the packages, and the advertising techniques and acumen (if not genius) needed to market them and assure what we would now call brand loyalty. In so doing, first women and then children were targeted as the natural consumers of sweets; by the mid-nineteenth century onward, sugar had become the currency of childhood.

Numerous industries would also contribute to our ever-increasing sugar consumption by using sugar in food preparation, but for reasons other than the sweetness itself. Flour milling was one of the many technological revolutions in the nineteenth century, for instance, and as the mills ground the flour ever more pure and white, even the yeast bugs saw little benefit from eating it. Sugar was added by the bakers to make the yeast rise, and rise faster, and to make palatable otherwise tasteless flour. Through the decades of the twentieth century, the sugar content in bread rose steadily, feeding what might have been an ever-more-demanding sweet tooth. (As *Sugar: A User's Guide* explained in 1990, white bread — the Wonder Bread of American childhoods, for example — can have a sugar content greater than 10 percent, compared with roughly 2 percent in European breads.)

Five industries in particular emerged beginning in the 1840s to contribute directly to the sugar saturation of our diets and our lives by producing and marketing foods and beverages in which sugar was the primary or defining ingredient. We can think of these foods and beverages as doing for sugar what cigarettes did for tobacco (and all of them would eventually be targeted to children). Fruit juices, sports drinks, and especially breakfast cereals would appear in the market and then explode in popularity a century later, in the decades following the Second World War.

Candy

In 1847, a Boston druggist named Oliver Chase launched the modern candy industry with his invention of a machine for churning out perfectly formed candied lozenges by the thousands. Hand-cranked machines like Chase's would later become horse-powered, then steam-powered, and eventually electric-powered; local hand-produced sweets for the rich became mass-produced wholesale treats for the nation. The confection shop — "a display of grown-up prestige," as the historian Wendy A. Woloson explained in *Refined Tastes* — turned into the candy shop, "a venue for the children of early American capitalism." By 1876, when the city of Philadelphia hosted the Centennial Exposition, twenty companies were displaying mass-produced candies, created by specialized machinery. By 1903, *The New York Times* was estimating yearly candy industry sales at $150 million in the United States alone, up from "almost nothing" a quarter century earlier.

CHOCOLATE

The chocolate bar also dates to the 1840s, when Swiss confectioners — the Lindt brothers — figured out the trick of solidifying chocolate powder into a bar that could be mass-produced, packaged, and shipped. Until then, chocolate had been consumed as a hot beverage; only high-end French confectioners had known the secret of making edible chocolate in solid form. By the end of the century, automated machines to wrap individual bars were operating in factories throughout the United States, and Milton Hershey, among others, had begun mixing the chocolate with milk to make it sweeter, more delicately flavored, and thus more appealing to children. A remarkable proportion of the chocolate staples of the twentieth century and today were first created and mass-produced between 1886 (the Clark bar) and the early 1930s — Tootsie Rolls (1896), Hershey's Milk Chocolate bar (1900), Hershey's Kisses (1906), Toblerone (1908), the Heath bar (1914), Oh Henry!

(1920), Baby Ruth (1921), Mounds and Milky Way (1923), Mr. Goodbar (1925), Milk Duds (1926), Reese's Peanut Butter Cups (1928), Snickers (1930), Tootsie Roll Pops (1931), and the Mars and 3 Musketeers bars (1932).

ICE CREAM

Ice cream had been a treat for the wealthy since it was first invented — apparently in Italy — in the late seventeenth century. By the mid-eighteenth century, it was still sufficiently rare in the United States that eating it was considered an event worthy of mention in the newspaper. What it required to go viral, other than suitably inexpensive sugar, was either a reliable supply of ice or a freezer in which to make and store it. The natural ice industry — harvesting ice from Northern lakes, ponds, and rivers in the winter and preserving it throughout the year — exploded in the nineteenth century. The first ice-cream freezer was invented in 1843 by a Philadelphia tinkerer named Nancy Johnson.

Wholesale ice-cream production began with Jacob Fussell, a Maryland milk-dealer, who found himself in the summer of 1851 with an oversupply of cream and no customers to buy it. He added sugar, froze it into ice cream, sold it for twenty-five cents a quart, and was overwhelmed with the demand. Fussell then went into the wholesale business, opening ice-cream factories first in Pennsylvania, near the source of the cream, then in Baltimore, near his clients, and then in Washington, Boston, and New York. In England, an Italian pastry-maker named Carlo Gatti first began mass-producing ice cream in the late 1850s.

Ice-cream making might have been the one culinary talent in which the United States led the world. By the 1870s, druggists were adding ice cream to the soda water they had been dispensing in their establishments for forty years* (first plain, and later with flavorings and sweeteners). The result, as Woloson says, was "not only a new treat — the ice cream soda — but also a new institution — the ice cream soda fountain." By 1892, the ice-cream sundae had been invented; in 1904, the ice-cream cone pioneered at the World's Fair in Saint Louis;† in 1919, the Eskimo Pie; in 1920,

* Soda water had been invented by Joseph Priestley in 1767.

† Among the several existing creation myths, one that is taken seriously is that Ernest Hamwi, a waffle maker, had a concession stand at the fair

the Good Humor bar; in 1923, Popsicles.

SOFT DRINKS

And then there was soda pop. Dr Pepper, Coca-Cola, and Pepsi were all launched in the 1880s. A late-twentieth-century Coca-Cola CEO would describe the latter two as "the magnificent competitors," dominating the industry and competing in the dissemination of their products — flavored, caffeinated sugar water — to every last backwater in the world.

Soft drinks began as variations on patent medicines, which would become a lucrative industry in the second half of the nineteenth century. Coca-Cola was the conception of John Pemberton, an Atlanta maker of patent medicines, whose revelation was to mix the formulation for Vin Mariani — an exceedingly popular French wine (among its fans were Thomas Edison, H. G. Wells, President William McKinley, and six French presidents), infused with the powdered leaves of the coca plant (cocaine) — with kola nuts, another popular ingredient in pat-

---

next to an ice-cream dealer who ran out of cups in which to sell his ice cream. Hamwi rolled his waffles into cones, the ice cream was added, and the rest is history.

ent medicines, and the carbonated water being dispensed in soda fountains. Pemberton removed the wine from his formula in 1885, when local counties in Georgia voted to ban the sale of alcohol. That's when he added sugar to disguise the natural bitterness of the kola and the coca leaves. He advertised the mixture as "a delicious, exhilarating, refreshing and invigorating Beverage . . . a valuable Brain Tonic, and a cure for all nervous affections — Sick Head-Ache, Neuralgia, Hysteria, Melancholy, etc."

In 1891, Pemberton sold the Coca-Cola rights for twenty-three hundred dollars to Asa Candler, a former drugstore clerk and another maker of patent medicines, who set about creating a distribution network that within four years would have the product available in soda fountains in every state in the country and, within another two, in Canada and Mexico. In 1902, with a national debate raging about the addictive nature of cocaine, Candler had it quietly removed from Coca-Cola. This didn't seem to put a dent in sales. Coca-Cola was by then spending a hundred thousand dollars a year on advertising. When John Candler, Asa's brother, was asked what items Coca-Cola used for advertising, he replied, "I don't know anything they *don't* advertise on." By 1913, the company had upped its advertising budget to over a million dollars yearly, promoting Coca-Cola on over one hundred million items, including thermometers, cardboard cutouts, matchbooks, blotters, and baseball cards. Pepsi-Cola (originally called "Brad's Drink") came along thirteen years after Coca-Cola and was, as the name now implied, a direct competitor, its growth curve exponential. Pepsi-Cola syrup sales increased tenfold between 1904 and 1907; by the end of 1908, Pepsi had licensed 250 bottlers in twenty-four states.

The only setback to the ever-increasing levels of sugar consumption worldwide was the First World War, and that setback was temporary. The war in Europe took a third of the world's sugar supply — the European and Russian beet-sugar industry — out of circulation. The Cuban and American industries upped their production capacity to make up the shortfall, as did sugar industries in nearly fifty other countries around the globe. Rationing during the war was replaced afterward by the greatest yearly increases in consumption the United States had ever seen. Only in Europe was sugar consumption slow in returning to prewar levels. "The people of Europe have lost their sweet tooth," as one sugar-industry executive opined to a *New York Times* reporter in 1921. "They learned to do without sugar during the war. They are still doing without it, to a large extent; some from necessity, some from choice. It will require an energetic campaign of education to bring Europe back to her former sugar consuming status."

By then, the sugar industry in the United States was selling annually more than a hundred pounds of sugar per capita for the first time in history, and Americans were consuming more than three billion bottles of soft drinks a year. Journalists, historians, and sugar-industry executives were marveling at what had been accomplished in the previous century in driving up both sugar production and consumption, and in changing the nature of the American food supply.

Chapter 3
The Marriage of Tobacco and Sugar

> Such an investigation is pertinent not only because the cigarette consumption has reached an all-time high in the United States, but the American blended cigarette, this product of the marriage of tobacco and sugar, is now rapidly gaining popularity all over the world.
>
> "Tobacco and Sugar"
> Sugar Research Foundation, Inc.,
> October 1950

This book is about the likely consequences to human health of consuming significant amounts of sugar — eating it or drinking it. But the industrial revolution led to another significant change in human habits in the first half of the twentieth century that has had demonstrable effects on our health — the explosive success and dissemination worldwide of the American blended-tobacco cigarette and, with it, as I've discussed, the epidemic of lung cancer that cigarette smoking demonstrably causes.

Just as diabetes was an exceedingly rare disease (or at least diagnosis) prior to the industrial revolution and the steep rise in sugar consumption that followed, lung cancer was an exceedingly rare disease until cigarettes surged in popularity and transformed an uncommon disease eventually into a scourge. Only 150 cases of lung cancer were diagnosed in the United States in total prior to 1900. In 1914, one year after R. J. Reynolds introduced Camels, the first brand of cigarettes to be made of multiple tobacco types blended together, and the first year that lung cancer was officially listed as a cause of death in the United States, four hundred cases were diagnosed. By 1930, that number had increased sevenfold. In 1945, more than twelve thousand Americans died of lung cancer. In 2005, when the epidemic may have peaked, more than 163,000 Americans succumbed to the disease.

A story that has been little told — although Robert Proctor of Stanford University tells it in *Golden Holocaust,* his monumental 2011 exposé of the cigarette industry — is that sugar played, and still does, an absolutely critical role in this epidemic.

Proctor relies for much of this history, as do I, on a 1950 report, "Sugar and Tobacco," generated for internal use by the sugar industry's Sugar Research Foundation (SRF).* "This business of sugar in tobacco leaf is a fascinating one," Proctor says, "and insufficiently appreciated outside the tobacco man's labs."

For those who would immediately dismiss the possibility that sugar itself may be responsible for more premature deaths than cigarettes, we have to consider the fact that cigarettes themselves would have been far less harmful and far less addictive had it not been for sugar. "Were it not for sugar," Wightman Garner, a former chief of the tobacco branch of the U.S. Department of Agriculture, told the author of the SRF report in 1950 (back when the USDA could still conceivably be proud of what the tobacco industry had accomplished), "the American blended cigarette and with it the tobacco industry of the United States would not have achieved such tremendous development as it did in the first half of this century."

* The report acknowledges contributions from dozens of researchers and administrators, many of them at the U.S. Department of Agriculture.

Until the early twentieth century, Americans mostly smoked cigars or pipes, rarely inhaling the smoke of either, or they chewed "plug" tobacco, as it was then called. Cigarettes only overtook cigars and pipes in the mid-1920s (as measured by pounds of tobacco consumed), in part spurred by the distribution of cigarettes to the millions of young American men who fought in the First World War, and in part by the ever-increasing popularity of American blended cigarettes. Within two years of its introduction by R. J. Reynolds, Camel was the best-selling cigarette in America; within eight years, Camel accounted for 40 percent of all cigarettes sold. By the 1930s, cigarette manufacturers in the United States were selling almost exclusively blended cigarettes, and the American blended cigarette was in the process of taking over the world — an accomplishment, as with Coca-Cola and Pepsi, that the Second World War would aid immeasurably.

The critical factor driving both addiction and cancer is that cigarette smoke can be easily inhaled. When tobacco is drawn deep into the lungs, the nicotine can be absorbed, along with oxygen itself, over an internal surface area that has been estimated to be roughly half the size of a tennis court. (At most, 5 percent of the nicotine in tobacco smoke is absorbed in the mouth, according to Wightman Garner's 1946 book, *The Production of Tobacco.* "When the smoke is inhaled, a much greater proportion of the nicotine is absorbed.") But this huge surface area also offers enormous opportunity for healthy cells to be targeted by carcinogens and transformed into malignant cells, and so what makes the experience of smoking cigarettes so pleasurable and so addictive — what gives the "nicotine satisfaction," as tobacco researchers would call it — is also critical to the cancer process as well. The cigarette industry could have made cigarettes that were harder to inhale, notes Proctor, and so the nicotine would have been less addictive, but then they'd have sold fewer cigarettes and hooked fewer smokers.

American blended cigarettes, as the name implies, are blends of multiple types of tobacco. The two most prominent tobaccos in blended cigarettes — about 70 percent of the content — are air-cured Kentucky or "Burley" tobacco, and flue-cured Virginia tobacco. It's flue curing that constituted the great technological revolution in the tobacco industry in the 1860s and 1870s, making inhalation possible, as Proctor tells it, and leading him to suggest that "flue-curing may well be the deadliest invention in the history of modern manufacturing. Gunpowder and nuclear weapons have killed far fewer people."

When tobacco is flue-cured, the harvested tobacco leaves are suspended over iron flues that heat the surrounding air to progressively higher temperatures. The process continues for the better part of a week, during which the heat first fixes the color of the tobacco leaves and then dries them, while breaking down the enzymes in the leaves that would otherwise break down the sugars they contain. Tobacco that begins with a relatively high carbohydrate content (up to 50 percent of dry weight) but is low in sugar (3 percent) ends up as much as 22 percent sugar, sucrose specifically. The "closest parallel" to what happens in the tobacco leaves during flue curing, notes the 1950 SRF report, is "the massive conversion of starch into sucrose" that happens when bananas are harvested and allowed to ripen.

The sugar content of the flue-cured tobacco leaves is the key to inhalation. The high sugar content results in tobacco smoke that is acidic rather than alkaline — chemists would say that it has a lower pH. Alkaline smoke irritates the mucous membranes and stimulates the coughing response. Acidic smoke can be inhaled without doing either. Most people, as German researchers noted in the 1930s, are unable to inhale the alkaline smoke from pipe and cigar tobaccos, but they can inhale the acidic smoke from the sugar-rich, flue-cured tobacco in cigarettes. So this is the first of two roles played by sugar in blended cigarettes that are critical to inhalation and addiction.

Until Camel came on the market, cigarettes were made almost exclusively from flue-cured tobacco. Though they could be inhaled, they had a relatively low nicotine content, and the nicotine was not easily absorbed by the lungs. The more sugar naturally occurring in the tobacco, the lower the nicotine content, and the less absorbable the nicotine is. As such, the satisfaction to be derived from the experience of smoking cigarettes prior to Camel was also low, at least compared with that of cigars or pipes or chewing plug tobacco, all of which used predominantly the air-cured Burley tobacco. A novice smoker's urge to keep smoking or to smoke frequently was also relatively low.

In 1911, the Supreme Court dissolved the American Tobacco Company — known as the Tobacco Trust — on the grounds that it was a monopoly and thus in violation of the Sherman Antitrust Act. In doing so, it split the trust into four smaller companies. One was R. J. Reynolds, which had sold chewing tobacco and now moved into the cigarette business. For its Camel cigarettes, R. J. Reynolds used a tobacco blended from the air-cured Burley of their chewing tobacco and the flue-cured Virginia tobacco traditionally used in cigarettes (as well as some sun-cured Oriental tobacco midway between Burley and Virginia tobacco in sugar content, and minor amounts of other tobaccos).

Air-curing Burley tobacco results in a tobacco that's relatively nicotine-rich, and the nicotine is easier to absorb than it is in Virginia tobacco, but the smoke itself is alkaline and thus difficult to inhale. More important, after air curing, Burley tobacco has virtually no sugar in it, which is what Wightman Garner described in 1946 as one of its "objectionable properties." But by 1913, this problem had been solved by the makers of plug tobacco, and the Burley tobacco that went into Camel was already what Proctor aptly described as a "candied up" tobacco.

The leaves of Burley tobacco are porous

and absorbent, a quality that prompted the earliest tobacco farmers in Missouri and Kentucky to realize that Burley leaves could easily absorb sugar. These tobacco farmers had taken to sweetening their tobacco after curing with a process that immersed the leaves in a "sugar sauce," marinating them, in effect, in a concentrated sugar solution that might also typically include honey, maple syrup, molasses, fruit syrups, licorice, and other sweeteners.* As the Sugar Research Foundation would point out, "Sugar enhances the flavor of aromatic substances, just as it does whenever it is applied in prepared and processed foodstuffs." Burley tobacco can absorb up to 50 percent of its own weight in sugar through the saucing process, and manufacturers of chewing tobacco took advantage both to make their products sweeter and to save money, because sugar, pound for pound, was cheaper than the tobacco. (Virginia tobacco farmers in the 1880s blamed competition from the

* When sweetened chewing tobacco was first commercially produced, in the 1830s, it sold with "sensational rapidity," as the Duke University historian Nannie May Tilley wrote in 1972, and the tobacco growers who pioneered the process "in a few years amassed a fortune."

sugar-sauced tobacco on "the perverted tastes of the Yankee who did not care for tobacco but dearly loved sweets.")

It was this sugar-sauced Burley tobacco that R. J. Reynolds blended into Camels, a decision that the SRF report called either an act of "necessity [they had mainly stocks of air-cured tobaccos used in the manufacture of plug] or the stroke of genius anticipating future trends in demand and consumption." Either way, if the explicit goal had been to maximize the delivery of nicotine — and so, regrettably, carcinogens with it — to the human lungs, they may not have been able to find a better way to do it. American cigarette manufacturers all followed suit.

By 1929, U.S. tobacco growers were saucing Burley tobacco with fifty million pounds of sugar a year and using it in over 120 billion cigarettes.* The sugar balanced out the tobacco's naturally alkaline smoke, maximizing its inhalability and delivering even more nicotine into the lungs. The sugars in

* By 1939, according to the Sugar Research Foundation report, 40 percent of all the maple sugar produced in the United States and "almost all" of the imports from Canada were being used to sauce tobacco.

the tobacco also "caramelize" as they burn (technically, during the process of pyrolysis) and the caramelization of the smoke provides a sweet flavor and an agreeable smell that made cigarettes more attractive to women smokers and to adolescents as well. ("This [caramelization] process adds as much to the flavor and smoking enjoyment of cigarettes as it does to the arena of confectionary and bakery products," notes the SRF report.)

Since the 1970s, toxicologists and cancer researchers have been studying the effect of sugars in cigarette smoke and confirming the observations made by the Sugar Research Foundation report in 1950. As toxicologists in the Netherlands explained in 2006, "Consumer acceptance of cigarette mainstream smoke [what's directly inhaled] is proportional to the sugar level of the tobacco." These researchers pointed out one other interesting if regrettable aspect of the acidic smoke that comes from the sugary tobacco used in cigarettes: The acidity of the smoke increases as the cigarette burns closer to the butt, as does what chemists call its "acid buffering capacity," which in turn decreases the absorbability of the nicotine. This means that as the cigarette burns down, the nicotine satisfaction decreases and the smoker tends to draw longer and harder to compensate. As a result, the urge to inhale most deeply is greatest when the tar-and-carcinogen content of the smoke is also greatest. The opposite is true with air-cured tobacco in cigars, in which the smoke becomes progressively more alkaline, thus increasing the absorbability of the nicotine and lessening the urge to inhale.

When the Sugar Research Foundation produced its report on sugar and tobacco in 1950, four years after Wightman Garner of the USDA confirmed the key role that sugar played in the explosive growth of the cigarette industry, neither had reason, or at least reason enough, to consider the deleterious consequences. Both were thinking of how the sugar industry could continue to benefit from the cigarette industry's remarkable growth. "This spectacular development," proclaimed the SRF report, "sets no limit for possible expansion of sugar use in tobacco products and especially cigarettes. While most of it will certainly depend on future demand for American-type blended cigarettes at home and abroad, there is also a possibility of using cane and beet sugar to a larger extent to make up for sugar deficiencies in tobacco types used in blended cigarettes." Fourteen years later, the surgeon general's landmark report on smoking and health would officially link cigarettes to lung cancer, giving the sugar industry reason to rethink this position. Still, as the SRF report correctly claimed, it was the "marriage of tobacco and sugar" that made possible both the astounding success of American cigarettes worldwide and the lung cancer epidemics that followed.

CHAPTER 4
A PECULIAR EVIL

In 1937, C. W. Barron, then the owner of *The Wall Street Journal,* made the pithy observation that if we want to make money in the stock market, we should invest in companies that provide us with our vices. "In hard times [consumers] will give up a lot of necessities," he said, "but the last thing they will give up is their vices."

George Orwell made a similar observation that same year in a very different context, documenting the bleak lives of the British laboring class in *The Road to Wigan Pier.* In a decade of unparalleled depression, Orwell observed, sales of what he called "cheap luxuries" had surged. "The peculiar evil is this," he wrote. "A millionaire may enjoy breakfasting off orange juice and Ryvita biscuits; an unemployed man doesn't. . . . When you are unemployed, which is to say when you are underfed, harassed, bored and miserable, you don't *want* to eat dull wholesome food. You want something a little bit 'tasty.' There is always some cheaply pleasant thing to tempt you."

This observation alone may be enough to explain the resiliency of the sugar industry, regardless of how hard the times, and of the "depression-proof" nature of candy, ice cream, and soft drinks. Annual per capita sugar consumption in the depth of the Great Depression was sixteen pounds *higher* than it had been in 1920. Candy consumption climbed steadily through the Depression. Coca-Cola thrived, as did Pepsi, although not before first declaring bankruptcy in 1931. An investor who purchased Coca-Cola stock at its highest price in the summer of 1929, held it through the Crash and the ensuing Depression, and then sold it in 1938 at its lowest price, as *Barron's* reported at the time, would have made a profit of 225 percent. It was during the Depression that Schrafft's restaurant chain in New York City reported diners "breakfasting on Coca-Cola and rolls or even Coca-Cola alone," rather than the more nourishing meals they might have eaten when they had the money.

Until the second-to-last year of the twentieth century, the one certainty about sugar was that consumption increased, if not every year, then over time. Sugar shares a common feature with those agricultural products for which the demand and supply are relatively immune to the price — what economists call "price inelastic." As the economists Stephen Marks and Keith Maskus have noted, rising prices don't lead to less consumption in these cases; they lead to greater production and eventually greater revenues for the producers. But falling prices also lead to greater demand and production. Production and consumption move steadily upward.

In the sugar industry, these cycles invariably begin with production shortfalls. For instance, storms or droughts in the tropics disrupt cane-sugar production; wars in Europe and Asia have disrupted beet-sugar production or restricted trade. Less sugar is available, and so prices rise. Reserve stocks are quickly depleted. The public demands more sugar. As Earl Babst, a president of the American Sugar Refining Company, said about the specter of sugar rationing during World War I, a "frantic and abnormal demand" resulted. Other producers around the world make up for the shortfall by planting more sugarcane or beets, building more sugar factories, and increasing refining capabilities to process that sugar. The more sugar these producers can grow, refine, and sell, the greater their profits.

Once the disrupted sugar fields come back on line, though, the supply of sugar exceeds the demand. And because sugarcane continues to produce sugar for half a dozen years after planting, the farmers will continue to harvest it until they have to pay more to harvest it than they can get from selling it. The refiners will refine it. The result is a post-disruption glut in available sugar, which causes prices to plummet. This was "the unhealthy economics and unholy politics," as *Time* magazine phrased it in 1945, which led to an industry that "produces too much sugar between wars and too little during them." Sugar growers and refiners are naturally resistant to the idea that they produce less to rein in prices; the sugar fields, whether beet or cane, are typically unfit for other crops that might be planted instead.

The industry invariably responds to the glut and plunging prices by lobbying governments for policies — import quotas and subsidies — that will protect producers from losing money, while allowing them to continue to harvest and process all the sugar they can. The industry will also work diligently to increase consumption globally, looking for new industrial uses for sugar, and promoting sugar directly to the public. This strategy includes inducing countries that import and consume little sugar — China, for instance, as was suggested in 1931 — to increase their consumption.

By the mid-1930s, when the U.S. Congress passed the Sugar Act, which would stay in force, with amendments, for forty years, the domestic sugar industry was distributed so widely — beet sugar in the Northern, Central, and Western states; cane in the South; refiners on the coasts; and the candy, soda, and paint industries (sugar is an essential ingredient in paint) — that President Franklin Roosevelt was calling the sugar lobby, according to *The New York Times,* "the most powerful pressure group that had descended on the national capital during his lifetime." The Sugar Act effectively guaranteed that producing and refining sugar in the United States would always be a profitable business. It established the price of raw sugar (typically higher, if not significantly so, than world prices), put limits on domestic production, and set quotas on imports. The Sugar Act also allowed for subsidies to be paid to producers either for the sugar they didn't produce or the sugar they couldn't sell —

"benefit payments to domestic producers," in the words of the *Times*. As a result, consumers were invariably paying more for sugar than would have been the case without the quotas and price supports. And yet that didn't stop us from buying sugar.

Technological advances continued to work to the benefit of the sugar industry. Sugar-rich products could be made ever more available to consumers. Vending machines — "electric coolers" — made their appearance in the 1930s, and the price of refrigerators dropped so much that they became common household appliances. By 1935, refrigerators could be purchased for well under two hundred dollars, and one and a half million were sold that year alone. For the first time in history, consumers could easily indulge in ice-cold soft drinks and ice cream without leaving their homes. Coca-Cola and Pepsi began selling their products in markets in six-packs and cartons for home use, and crafting advertising campaigns that targeted women and children specifically. In the six years leading up to America's entry into the Second World War, soft-drink sales in the United States nearly quadrupled — from two hundred million to 750 million cases per year.

The war created a setback but, as with the

First World War, only a temporary one. Sugar rationing began in 1942, with the Asian, European, and South Pacific industries no longer providing sugar to the West, and molasses in the United States being diverted to make industrial alcohol for the war effort (for synthetic rubber and explosives, primarily). A hurricane and a drought in Cuba disrupted the Cuban sugarcane industry, on which the United States relied for much of the sugar it consumed. By 1945, American civilians were expected to get by on levels of sugar consumption that hadn't been seen since the 1870s — only seventy pounds per year. One economist was calling it the "worst sugar famine in history."

The dearth of sugar available for civilian use was compounded by the massive allotment of sugar going to the eleven million servicemen of the armed forces — 220 pounds per capita yearly for the U.S. Army, according to a 1945 congressional investigation. This was twice what the soldiers would have been eating prewar as civilians, and more than three times the amount allotted to noncombatants on the home front. It seemed excessive even to the congressional investigators, but they wouldn't interfere, lest they be seen as harming the war effort.

"It would not seem unreasonable," the committee suggested, "for some responsible officer of the American armed forces to inform all area commanders of the stringency of the civilian sugar situation and ask their cooperation to conserve sugar in every way possible."

Toward the end of the war, authorities were touting the value of sugar and candy as stimulants to make "our warriors . . . more effective in combat," and the army alone was purchasing over a hundred million pounds of candy a year for its troops. Both the K-ration and the emergency D-ration had contained chocolate bars; the former included "fruit candy" bars as well. According to one navy analysis, candy bars constituted 40 percent of the foods that servicemen were purchasing from the mess over and above their sugar-rich rations. "We have tended to underestimate the importance of these bars in the feeding of men," reported the Cornell University nutritionist Clive McCay, who served as a commander at the Naval Medical Research Institute during the war years. The candy industry promptly took advantage of all this by launching an advertising campaign touting candy on the basis of its "fighting food value." The goal, as *The New York Times* suggested, was "to correct popular misinformation that candy is fattening and causes tooth decay."

Coca-Cola and Pepsi both made their service to the war effort the easy availability of their products to servicemen worldwide. Pepsi circumvented the rationing problem by stockpiling sugar at the start of the war and then importing syrup directly from Mexico as the war continued. The company set up Pepsi-Cola centers for servicemen that stayed open past midnight and served two million men in the first year of operation.

Coca-Cola won an exemption from the sugar rationing for Cokes sold to the military. The official Coca-Cola policy was to sell servicemen Coke anywhere in the world for a nickel a bottle, regardless of the cost to the company. To help accomplish this task, and to prepare for the postwar years, the company established sixty-four bottling plants worldwide, some using German and Japanese prisoners of war to work the plants. The company's unpublished history credited this policy with making "friends and customers for home consumption of 11,000,000 GIs" and doing a "sampling and expansion job abroad which would [otherwise] have taken 25 years and millions of dollars."* When the company hosted its first international convention three years after the war ended, one of its executives described its purpose as the beginning of the effort necessary to "serve those two billion customers who are only waiting for us to bring our product to them." "When we think of Communism," read a sign at the conference, "we think of the Iron Curtain. BUT when THEY think of democracy, they think of Coca-Cola."

When *Time* magazine put Coca-Cola on the cover in 1950 — with the Coke symbol lovingly feeding a Coca-Cola to a thirsty globe — a third of the company's profits were already derived from international sales. And Pepsi, of course, was quickly catching up: Its sales abroad increased fivefold in the 1950s, as the company opened two hundred bottling plants outside the United States. By 1959, Vice President Richard Nixon would be photographed in Moscow with Soviet Premier Nikita Khrushchev, both holding bottles of Pepsi.

While sugar consumption was rebounding

* After the war, one Coca-Cola employee working in Eastern Europe observed that Coke was second only to Hershey bars as an inducement for sex with the local women.

in the postwar years, the ways in which we consumed it once again shifted. Soft drinks, candy, and ice-cream sales would regularly hit new highs — ice-cream consumption alone doubled between 1940 and 1956 — but now sugar would become a mainstay of breakfasts as well, first in fruit juices and then in sugar-rich breakfast cereals.

Canned breakfast juices had first appeared during Prohibition, motivated by grape growers who could no longer sell their products as wine, and by orange growers in California and Florida burdened with surplus oranges during years of glut. In 1920, a cooperative of California growers (selling under the now familiar brand name Sunkist) began taking advantage of what nutritionists of the era called the "new nutrition" — the awareness of the importance of vitamins in preventing deficiency diseases — and took to advertising their products as a healthy way to get necessary vitamins, particularly vitamin C, a proposition that's still with us today.

Many consumers had become accustomed to drinking fruit juices instead of alcohol during the Depression. The "crowning achievement" in fruit-juice history, however, according to *The Oxford Encyclopedia of Food and Drink in America,* and "perhaps a

defining moment of the American breakfast," was the invention of frozen concentrate by researchers funded by the federal government in the years after World War II.

Minute Maid, in 1948, was the first. By the mid-1950s, "chilled" orange juice had also arrived. By 1980, according to USDA estimates, Americans were drinking over seven and a half gallons of fruit juice a year, and by the late 1990s, when the trend (as with sugar consumption itself) peaked, over nine gallons — roughly equivalent to drinking an additional eight pounds of sugar per year. These sugar-rich juices would not show up in the official USDA estimates of sugar consumption.

Fruit juices could easily be marketed, as the fruit industry did, as healthful additions to the American diet, and company nutritionists would go along. This was not the case with breakfast cereals, which further transformed American breakfasts in the 1950s. The company nutritionists had second thoughts. They were able to delay the appearance of sugar-coated cereals for perhaps half a century, and then market forces overwhelmed them. By the 1960s, children's breakfasts had been reshaped into a morning variation on the theme of candy bars or dessert — perhaps lower in fat content, but richer than ever in sugar. Companies would offer all sorts of rationalizations for the creation of cereals that in some cases were over 50 percent sugar, and they would market them relentlessly to children. Once a single cereal company broke through the pre-sweetened barrier, the others did it — or so they told themselves — to survive.

The dried-cereal industry had its roots in Battle Creek, Michigan, and the health-food movement of the late nineteenth century. The pioneers were John Harvey Kellogg, a physician who was a follower of the Seventh-day Adventist Church, and his competitor and former patient, C. W. Post. Both operated what they called "sanitoriums" for the well-heeled dyspeptic,* and both believed that the path to health and happiness ran through the digestive tract. As Kellogg would say, "The causes of indigestion are responsible for more deaths than all other causes combined." The idea of a breakfast flake that would aid digestion supposedly came to Kellogg in a midnight revelation, and he set to work on it the following morning. Post beat him to it, though, with his Grape Nuts, which by 1900 had earned him what was then the single largest, fastest legitimate fortune in America.

* Kellogg's many famous patients included J. C. Penney, Montgomery Ward, John D. Rockefeller, Eleanor Roosevelt, and Johnny Weismuller.

Post Grape Nuts were originally made with molasses and maltose from barley flour, but no cane or beet sugar. Kellogg's first cornflakes were sugar-free as well. But Kellogg had put his younger brother, W.K., in charge of the development progress, and while the elder Kellogg was away in Europe in 1902, W.K. added sugar to the toasted cornflakes to improve the taste and the flaking process. John Harvey was said to be outraged when he returned — "he felt that sugar was unhealthy and argued vehemently against using it," as the story is told in the 1995 history *Cerealizing America.* Consumers disagreed, though, and the sugar — a relatively trivial amount — stayed. Two years later, when Quaker Oats gave away a truly sugar-coated cereal at the 1904 World's Fair in St. Louis, the company considered it candy, as did their customers, and chose not to market it, on the assumption that "America's sweet tooth was a passing fad." This turned out to be not quite correct.

It took thirty-five years for dried cereals, a health food, to begin the successful transformation into sugar-coated cereals, a hugely profitable breakfast candy. The process began with an industry outsider — Jim Rex, a Philadelphia heating-equipment salesman — and a line of thinking that seems almost incomprehensible in the context of the anti-sugar sentiments of today. As told in *Cerealizing America,* Rex was sitting at breakfast one day watching his children ladle spoonfuls of sugar atop their puffed-wheat cereal. "Sickened by the sugary excess, Rex began to think of ways he could get his kids to eat their cereal without plunging into the sugar bowl. The solution came to him in a flash of inspiration. Why not create a cereal 'already sugar'd.' "

The result, Ranger Joe, was the first sugar-coated, pre-sweetened cereal sold in America. Rex sold it in local markets, but he failed to solve the technical issue of the cereal's clumping together in its package because of the sugar coating — it would "turn into bricks," as one cereal-industry executive later put it. After just nine months on the market, Rex sold his company to another local entrepreneur, who in turn sold out in 1949 to the National Biscuit Company (now Nabisco). By then, Post Cereals was already planning to roll out a competitor, Sugar Crisp, nationwide.

Post then began the trend of rationalizing how a company positioned as a producer of health foods could justify selling a cereal coated in sugar. Echoing the logic of Jim Rex, Post executives would argue that pre-sweetened cereal actually contained less sugar than what children would add on their own. By adding sugar, Post was merely "trading off sugar carbohydrates for grain carbohydrates and sugar and starch are metabolized in exactly the same way." Biochemists had already demonstrated that this was untrue, but it was not widely known. Either way, Post argued that "the nutritional value of the product" remained unchanged, with sugar calories replacing those from cereal grains. Sugar Crisp (now called Golden Crisp) sold spectacularly well, forcing the rest of the industry to play catch-up. Nabisco quickly released Ranger Joe nationwide, now renamed Wheat and Rice Honeys. Kellogg's, in 1950, released Sugar Corn Pops, even though most of the company stock was still held by the W. K. Kellogg Foundation, "a charitable organization established to promote children's health and education."

Kellogg's set out to produce a sugar-coated version of its iconic cornflakes as if "it was their salvation," releasing Sugar Frosted Flakes in 1952 and Sugar Smacks, a direct competitor to Post's Sugar Crisp, a year later. Kellogg's failed to produce a sugar-coated oat cereal and turned to chocolate instead. The company logic, again guided by nutritionists, was that "all this sweetness is not the best for children, [and] that bittersweet chocolate was good and healthy and it wouldn't be harmful to them." The result was Cocoa Krispies. When the first, bittersweet-flavored version didn't sell, the company added even more sugar. "The new cereal," as one Kellogg's salesman put it, "was a dietary flop, and a sales bonanza."

General Mills executives worried about the "possible dietary effects" of sugar-coated cereals, and its in-house nutritionist delayed the company's entry into the pre-sweetened market for years, but eventually they were overruled. The marketing team at General Mills argued that if the company didn't compete, it wouldn't survive. In 1953, General Mills released Sugar Smiles, a mixture of Wheaties and sugar-frosted Kix; by 1956, they had released three more sugar-coated cereals — Sugar Jets, Trix, and Cocoa Puffs.

Over the next twenty years, the cereal industry would create dozens of sugar-coated cereals, some with half their calories derived from sugar. The greatest advertising minds in the country would not only create animated characters to sell the cereals to children — Tony the Tiger, Mr. MaGoo,

Huckleberry Hound and Yogi Bear, Sugar Bear and Linus the Lionhearted, the Flintstones, Rocky and Bullwinkle — but give them entire Saturday-morning television shows dedicated to the task of doing so.

These companies would spend enormous sums marketing each cereal — six hundred million dollars total in a single year by the late 1960s, when the consumer advocate Ralph Nader took on the industry. Each new cereal that succeeded would spawn a rush of imitators, while the industry, by the 1960s, was now openly advertising the candylike nature of the products: "It tastes like maple sugar candy," Marky Maypo's father said of Maypo in 1956, to entice his son to eat it; Cocoa Krispies were advertised as tasting "like a chocolate milk shake, only crunchy." Industry executives, bolstered by nutritionists — most famously, Fred Stare, founder and director of the nutrition department at Harvard — would justify the sale of sugar-coated cereals as a means to get kids to drink milk, or as part of a "healthy breakfast." The magazine *Consumer Reports* may have captured this logic perfectly in 1986 when it claimed, "Eating any of the cereals would certainly provide better nutrition than eating no breakfast at all."

The identical logic is still used today,

when nutritionists and public-health authorities argue that children should be allowed to drink sugary chocolate milk because the benefit of obtaining the vitamins and minerals in the milk outweighs any danger that could come from drinking the sugar. This is based on a conception of nutrition science that dates back to the "new nutrition" of the 1920s, and whether it is true or not, or even vaguely true, was and still is the obvious question.

CHAPTER 5
THE EARLY (BAD) SCIENCE

> In spite of the doctors, we declare that when sugars are dear the people suffer. When we are all obliged to deny the many little gratifications of our whimsical palates, we are made very uncomfortable.
>
> *The New York Times,* 1856

> Most people know that the sugars are good food. Some people know how many calories there are in a piece of fudge. A few people know that sugar is not conducive to reducing.
>
> J. J. WILLAMAN,
> University of Minnesota, 1928

By the early decades of the twentieth century, in medical journals and in newspapers, physicians could be found blaming sugar for a host of ills that seemed to come about with the dramatic increase in the product's consumption. Diabetes would get the most attention, as awareness spread of an apparent diabetes epidemic. Rheumatism, gallstones, jaundice, liver disease, inflammation, gaseous indigestion, sleeplessness, tooth decay, ulcers and intestinal diseases, neurological disorders (or at least "nervous instability"), cancer, and "making the human race a degenerate people" were all blamed on sugar, and for an obvious reason. "No other element in the human dietary has increased with such leaps and bounds," wrote the Los Angeles physician Alexander Gibson in *The Medical Summary* in 1917. "The prodigious feeders of the Elizabethan era, when sugar cost a guinea a pound, consumed less free sugar in a month than a modern school child for a couple of penny's worth of 'all-day-suckers' consumes in a day. In fact the indulgence of sugar has exceeded every other stimulant, even including tobacco, coffee, tea and alcohol."

Discussions on the value of sugar, the risks and benefits of consuming it in quantity, were informed by the science of nutrition, which was in its infancy. Typically, science makes progress when new technologies are invented or applied, allowing researchers to obtain new information, and thus to ask and answer new questions about the phenomena they're studying. In nutrition and its relationship to chronic disease, however, this never happened. New technologies appeared, and they resulted in new revelations, as expected, but those revelations had no influence on how nutritionists, and even researchers studying obesity and diabetes, perceived the problem presented by sugar. The thinking of the 1920s remained firmly set, and we've been living with the consequences ever since. Understanding how and why this happened is critical to understanding the risks and benefits of consuming sugar.

The roots of the modern science of nutrition date back to France in the late eighteenth century, and they coincide with the birth of modern chemistry, as a handful of now legendary scientists began to explore the relationship between the air we breathe, the foods we eat, and what it means, in effect, to be alive — the chemical reactions that constitute life itself. As the science of nutrition diverged from chemistry in the latter half of the nineteenth century, the nexus of research moved to Germany, where the details of how organisms burn protein, fat, and carbohydrates for fuel were worked out. ("The amount of information [the Germans] acquired within a comparatively few years past is remarkable," wrote the American nutritionist Wilbur Atwater in 1888.) Scientists there would study the metabolism and respiration of men and animals under various dietary conditions, studying the balance of energy into and out of the human body — what went in via breathing and eating, and what exited in the breath and as heat or excreta.

These were the obvious first questions to ask, and the tools the scientists had available drove their research — as is always the case in science. Historians would later date the birth of *modern* nutrition science to the 1860s, when German researchers pioneered the use of room-sized devices called calorimeters that allowed them to measure precisely how much energy human or animal subjects expended under different conditions of diet and physical activity. By the early twentieth century, nutrition researchers were measuring the energy requirements of children, soldiers, and athletes; they were studying how foods contributed to building strong bodies, and the components of a healthy diet — how many calories were needed, how much protein, and what vitamins and minerals. They studied what happened when essential vitamins and minerals were absent from the diet and identified deficiency diseases that could be cured by adding them back. This was the "new nutrition" of the era, and it has been the foundation of nutrition wisdom ever since.

However, when physicians and public-health authorities started questioning the effects of various carbohydrates and sugars on human health, this research could tell them precious little about anything other than energy metabolism. The influence of foods on what were then called "internal secretions" — on hormones such as insulin and growth hormone — was unknown, as was the influence on any pathological conditions, other than those that were caused by vitamin or mineral deficiencies. These subjects had yet to be studied.

Not until 1960 would researchers publish the details of a technique called the radioimmunoassay, which allowed the measurement of hormone levels in the circulation with accuracy, and in turn gave birth to the modern era of endocrinology — the study of hormones and hormone-related diseases. As a result, nutritionists had a ninety-year head start in thinking about diet in terms of its effect on "energy balance" — on the energy consumed and expended by the human body — rather than on the internal secretions, the hormones, that regulate such fundamental properties as how much fat we accumulate in our cells and the "partitioning" or "allocation" of the fuels we consume, whether we store them as fat, carbohydrate (glycogen), or protein, or burn them for fuel.

That ninety-year head start would be critical in establishing how nutritionists and medical researchers interpreted the risk/benefit ratio of consuming sugars, and it still affects how they think about these issues today. When nutritionists say that sugar is "empty calories," they're defining the problem posed by sugar in the science of the early twentieth century — in terms of the amount of energy (calories) and vitamins and minerals (empty) they contain — and ignoring the research, and an entire field of medical science, that came after. Those physicians, like Elliott Joslin, who did think about the influence of hormones on disease states — insulin, in particular, on diabetes — had little or no understanding of how foods influenced those hormones. That was the purview of nutritionists, and the nutritionists lacked the tools or, frankly, the awareness to pay attention.

Nutrition researchers of the late nineteenth and early twentieth centuries were beginning to understand that sugar had properties that set it apart from other carbohydrates, but they didn't understand the extent of those properties beyond the realm of energy and vitamin and mineral content, or why they might be relevant to obesity, diabetes, or any related disease. The chemists and nutritionists who studied the metabolism of these carbohydrates in the laboratory or in lab animals weren't doctors, and they weren't treating patients or thinking about the public-health implications of their work. The American physicians treating obesity and diabetes were not applying the skeptical and rigorous thinking of science, and yet it was their opinions that would forge the conventional thinking about the relationship between sugar and disease.

At a time when physicians in America were first confronting this rising tide of diabetic patients, medicine and science had little connection in the U.S., though that began to change in 1893, with the founding of the Johns Hopkins Medical School. Physicians interested in scientific research would travel to Europe to learn from the authorities there, as Joslin did, but medical schools themselves did not require physicians to study science or even to understand it. As late as 1900, only a single medical school in the United States — Johns Hopkins — required that applicants have a college degree. Many schools, according to a 1910 Carnegie Foundation report on the state of American medical education, did not even require that their students have finished four years of high school. Their primary criterion for acceptance was the ability and willingness to pay tuition. None of these medical schools supported research. In 1871, when Henry Percival Bowditch of Harvard set up what may have been the first academic laboratory in the country to pursue experimental medicine, it was located in an attic, and Bowditch's father paid for some of the equipment. Americans of this era were transforming the worlds of engineering and industry, but not medical science.

European researchers and clinicians pioneered all the fields of science relevant to understanding both obesity and diabetes — including nutrition, metabolism, endocrinology, and genetics — and dominated this research through the Second World War. These Europeans would come to radically different conclusions about the genesis of obesity and thus, by implication, diabetes as well, but the European research communities evaporated with the war, and these

European conceptions evaporated with it. European scientists would later write, as the Nobel Prize–winning physician and biochemist Hans Krebs did in 1967, about the need for centers of excellence in science, where young researchers could do an apprenticeship, learning literally at the bench of great scientists, who in turn had learned their skills and how to think critically from the bench of other great scientists. As Krebs wrote, "Scientists are not so much born but made." This culture of science, and these centers of excellence, were unfortunately absent in medicine in the United States, so American physicians who pursued scientific investigations were making it up as they went along, for better or for worse.

The dilemma posed by sugar is a clear one, or at least it is in retrospect. It had been delineated more than two thousand years ago, when Hindu physicians noted that sugar "promotes nutrition *and* [my italics] corpulency." That it has rather remarkable nutritional qualities, nutritionists would later come to accept as a given. Its history suggests it has medicinal qualities as well. But do those who get fat do so, as some suggested, through merely consuming sugar in excessive quantities, or through some unique characteristic of sugar itself?

The roots of the modern discussion on sugar and disease can be traced to the early 1670s, when sugar first began flowing into England from its Caribbean colonies (and this, of course, may not be a coincidence) and the habit of drinking sugared tea was becoming common. Thomas Willis, medical adviser to the duke of York and King Charles II, noted an increase in the prevalence of diabetes in the affluent patients of his practice. "The pissing evil," he called it, and became the first European physician to diagnose the sweet taste of diabetic urine — "wonderfully sweet like sugar or hon[e]y." It was Willis who appended the term "mellitus" ("from honey") to the name of the disease.* Willis attributed the diabetes he

* Willis's testimony stands as an exception to the observation that diabetes was an exceedingly rare disease prior to the twentieth century. In his posthumous discourse, *Diabetes or the Pissing Evil,* Willis wrote, "We meet with examples and instance enough, I may say daily, of this disease." This could be an exaggeration, as Robert Tattersall, a retired professor of clinical diabetes at the University of Nottingham in the U.K. and author of *Diabetes: The Biography,* suggests. It could be a reflection of the fact that Willis's patients were

was seeing among his wealthy London patients to "an ill manner of living, and chiefly an assiduous and immoderate drinking of Cider, Beer, or sharp Wines." But he nonetheless strongly "disapprove[d] [of] things preserv'd, or very much season'd with Sugar . . . [and judged] the invention of it, and its immoderate use to have very much contributed to the vast increase of Scurvy in this late Age."

Willis's denunciation of sugar led in turn to its censure by the botanist John Ray, which could "frighten the Credulous," as the physician Fred Slare noted in 1715, forty years later. (Scientific debates moved far more slowly in the pre-Internet era.) It was Slare's vigorous defense of sugar — his "Vindication of Sugars Against the Charge of Dr. Willis, Other Physicians, and Common Prejudices" — that would once again capture perfectly the dilemma posed by sugar and the framing of the debates to come.

To "defraud" infants of sugar "is a very cruel Thing, if not a crying Sin," Slare wrote, before discussing the anecdotal experience of those, like his grandfather,

---

wealthy and royalty, and thus most likely to be afflicted.

who lived to be a hundred, and the duke of Beaufort, who died at seventy-one, both of whom ate excessive sugar by the standards of the era (Beaufort, apparently, for any era — a pound daily for forty years).* Slare also recounted his own experience as edifying: he was "near Sixty-seven" and in excellent health, he wrote, while indulging in large quantities of sugar. "I write without Spectacles, and can read a small Print: can walk ten or fifteen Miles with Ease, and can ride thirty or forty Mile a day." More important, perhaps, he had outlived some eighty of his colleagues in the Royal College of Physicians, many of whom "were bitter enemies" of sugar. (This kind of argument — akin to saying my uncle Max smoked two packs of cigarettes a day and lived to be a hundred, ergo cigarettes do not cause lung cancer — would also be common in the sugar debates ever after.)

* Slare found it notable that the duke of Beaufort's internal organs, upon autopsy, were in excellent shape, and he still had his own teeth. The duke apparently believed a common adage: "That which preserves Apples and Plums, Will also preserve Liver and Lungs." Slare considered the duke's viscera and teeth to be evidence that the duke was right.

Slare also noted that "the worst of the Skum and Sediment" from the sugar refineries in the West Indies was used successfully to fatten hogs — a good thing, from Slare's perspective. He added a single caveat to his absolution of sugar as a dietary evil. Writing at a time when sugar was still a luxury item and its yearly consumption in England is estimated to have been less than five pounds per capita, or less than one-twentieth what it would be two centuries later, he nevertheless cautioned that women who prided themselves on their "fine proportions" but were "inclining to be too fat" might want to avoid sugar, because it is "so very high a Nourisher, may dispose them to be fatter than they desire to be."

Still, in an era when malnutrition and undernutrition were pervasive problems throughout Europe, sugar's ability to put fat on the lean or emaciated was widely perceived as one of its beneficial qualities. Not only could the aged live for many years on "scarcely anything but sugar," as the British physician Benjamin Moseley noted in his 1799 treatise on the subject, but "taken in tea, milk, and beer, [sugar] has caused lean people to grow fat, and has increased the vigour of their bodies." It may have been Moseley, having spent eighteen years working in the West Indies, who first suggested that slaves grow fat sucking on the juice of sugarcane during the harvest, an observation that would be repeated in medical writing through the early twentieth century. Not only could the juice from sugarcane bring health to the sickly, worm-ridden infants of slaves, Moseley wrote — "Give a negro infant a piece of sugar cane to suck, and the impoverished milk of his mother is tasteless to him" — but it did the same for adults as well. "I have often seen old, scabby, wasted negroes, crawl from the *hot-houses,* apparently half dead, in crop-time; and by sucking canes all day long, they have soon become strong, fat, and sleaky."

In 1865, Abel Jordão, a professor at the Medical School of Lisbon and a leading European authority on diabetes, suggested that this ability of sugar to put fat on the lean might explain the association between obesity and diabetes. Whereas most physicians, including most notably Joslin, would come to think that obesity caused diabetes, Jordão proposed that a kind of pre-diabetic state, caused by consuming too much sugar, could in turn cause obesity. If animals were fattened by being given sugars and starches, he reasoned, then it made sense that humans got fat when they had too much sugar in their circulation, which was the case in diabetes. "A robust adipose constitution is not a cause, but an effect of the complaint," Jordão explained. "I have seen some cases of lean individuals attacked with diabetes, who commenced to fatten." When Charles Brigham, then a medical student at Harvard and later a renowned surgeon, wrote an award-winning thesis on diabetes that was published in 1868, he expanded on Jordão's thinking and echoed Slare's caveat as well, but now from the opposite perspective: "On this same principle of sugar fattening," Brigham wrote, "many of the fairer sex, ashamed of the skeleton-like appearance which their shoulders and arms present when exposed, are in the habit of taking frequently a glass of eau sucrée [sugar water] in hopes of an amendment."

The few nutrition researchers and food chemists studying sugar and other carbohydrates were focusing their attention almost exclusively on sugar's nutritional qualities, determined solely by what they could measure at the time. By 1900, they had delineated the different types of sugars found in nature — glucose and fructose, for instance, which were then known as dextrose and levulose respectively — and the ways in which they combined in the more complex sugars, such as the lactose in milk, or sucrose from beet and cane. Researchers would report that muscles use these sugars for fuel and do so very efficiently. (They, too, would often, if not typically, confuse the sugar we consume — sucrose, composed of fructose and glucose — with the glucose of blood sugar.) Unlike protein, which leaves behind nitrogen to be excreted in the urine, carbohydrates produce energy "without any waste and leaving no residue." And although carbohydrates don't work to build muscle, as protein does, the body appears to burn them preferentially as fuel, sparing the protein in the process.

In 1916, Harold Higgins, working at the Carnegie Institute of Washington (located in Boston), measured how quickly our bodies metabolize these different sugars — how quickly, in effect, they give us energy; this was considered to be the "nutritive value" of the food. Higgins reported that we metabolize fructose and sucrose more quickly than other sugars. This finding would be the biochemical basis of the idea that sugar provides "quick energy," as the sugar industry would later advertise.

Higgins's laboratory research also confirmed the observation that sugar had what the British physician Willoughby Gardner, writing in the *British Medical Journal* in 1901, would call "unexpected stimulating properties." This observation distinguished sugar from other carbohydrates and suggested that it was, literally, a stimulant — the late-nineteenth- and early-twentieth-century version of a performance-enhancing drug. German researchers, wrote Gardner, had tested "various men, both of weak and of strong muscular physique," and concluded that an ounce of sugar was sufficient to restore within forty-five minutes "the power of work to muscles so tired that they had previously given hardly appreciable results." Sugar seemed to help these men perform "extraordinary muscular labor," and the Germans speculated that it might directly influence the nervous system to "overcome the feeling of fatigue."

Other researchers noticed similar effects in their experiments, and these observations supported reports from the field that lumberjacks, Alpine climbers, and polar explorers had taken to using sugar instead of brandy or other alcohol to relieve fatigue. Parisian cab companies had even taken to feeding sugar to their horses to give them energy and restore vitality. The legendary British climber George Mallory said that in his 1923 attempt on Mount Everest, he succeeded in making it within two thousand feet of the summit by living on sugar for the last few days of the ascent: almost exclusively lemon drops, peppermint candies, and chocolate. "At great elevations no one has any strength to waste on unnecessary processes of digestion," Mallory said; "sugar . . . can be digested quickly and easily converted into muscular energy. It has also a much-needed stimulating effect."

In 1897, according to Gardner, the German Reichstag had debated the value of sugar as a food and made the decision to test it on German soldiers, a trial that was carried out during autumn maneuvers the following year. "The results were conclusively in favor of the sugar eaters," Gardner wrote. The soldiers given sugar in their rations increased in weight, "which their comrades did not, they enjoyed better health, and were able to support the hard work with much less distress. . . . As a result of these experiments it was resolved that the sugar ration for the German soldiers should be raised to 60 grams per day." (That this happened to be almost twice what British soldiers were getting — thirty-seven grams — seemed to suggest to Gardner that the British were now at a distinct military disadvantage.)

Dutch authorities took to advocating "sugar training" for endurance athletes, and several rowing clubs — including the Rowing Society of Berlin — took up the practice of eating what were then considered large quantities of sugar and by doing so "did not become 'stale' or overtrained." By the mid-1920s, an era when rowing regattas were as popular as professional baseball or any other sport, rowing coaches at Harvard and Yale were emulating the Europeans and testing sugar on their rowers — jams, jellies, lumps of sugar, even a "pound of peppermints" (a "preposterous" rumor, suggested the Harvard coach: such an amount "would make a boy sick").*

In 1925, Harvard researchers reported in *The Journal of the American Medical Association* that runners in the Boston Marathon had very low blood sugar at the end of the race — similar to a diabetic, they wrote,

* In November 1924, the Yale soccer team was given sugar "in an attempt to increase their physical energy" during a game against the University of Pennsylvania. Yale lost, five to one, prompting a Yale professor of applied physiology to tell *The New York Times* that the results of the experiment "were noticeable but not convincing."

who is given "an overdose of insulin" — and that they had ameliorated the symptoms in other runners by having them load up with carbohydrates before the race and eat "glucose candies" while they ran, and supplying them with "tea containing a large amount of sugar at stations along the course." This report prompted editors at *The Lancet,* a British journal, to poke fun at the Americans for not knowing what everyone else had learned years earlier: "The most curious thing is that neither the authors nor the subjects at Harvard seem to have been aware that the consumption of sugar in one form or another is very widely known as preventive and curative of fatigue. . . . Sugar cakes are a sine qua non at an athletic tea-party."

Viewed from this quick-energy/fatigue-beating perspective, sugar seemed to be so valuable an item of the diet that the U.S. Department of Agriculture suggested that sugar "would seem to be a food especially adapted to children because of their great activity." By this logic, as Gardner suggested in the *British Medical Journal,* "the popular prejudice against" sugar was working to the detriment of growing boys and girls, not to their benefit. The candy industry, not surprisingly, agreed.

Through the 1920s, these discussions of sugar's nutritional value continued to be accompanied with what was usually an aside, that sugar was fattening and therefore the obese — anyone, for that matter, who had to work to remain lean — would be best served by avoiding it. As Gardner wrote in his assessment in the *British Medical Journal,* sugar was surely "one of the most valuable articles of the diet," and yet to be avoided "like poison" by those prone to obesity, diabetes, or gout.

This had become conventional thinking. After the artificial sweetener saccharin was discovered in coal-tar derivatives by Johns Hopkins University chemists in 1878, and then transformed into a commercial product over the next decade, it was immediately clear to medical authorities that "it may with benefit wholly or partially replace sugar in the diet" for the obese and diabetic, and perhaps those with liver disease and gout as well. In 1929, when delegates to the League of Nations met in Geneva to discuss economic issues facing their countries, one of the issues was the deleterious effect on their national sugar industries of "a growing world-wide abstinence by women" who were avoiding sugar "in order to keep their figures trim." By then, the American Cigarette Company was selling Lucky Strike — which began its existence as sugar-sauced plug tobacco and would beat out Camel in 1930 to become the nation's most popular cigarette — as "a splendid alternative to fattening sweets."

With the slowly rising tide of diabetes in the late nineteenth century, physicians and public-health authorities began entertaining the possibility that sugar was responsible. But because the disease was still relatively rare, so were the physicians who specialized in treating it and thought in a meaningful way about its cause. Elliott Joslin was among the first in the United States to specialize in diabetes, and he was just starting his career at the time. Joslin was followed by Frederick Allen, who had done research on diabetic animals at Harvard Medical School and on human patients at the Rockefeller Institute for Medical Research.

In 1913, Allen published a textbook on diabetes — *Studies Concerning Glycosuria and Diabetes** — compiling observations from human and animal studies, from the

* "Glycosuria" means an excess of sugar (glucose) in the urine.

biochemists, and even from history books. Allen's textbook included a lengthy discussion on the possibility that diabetes was caused by sugar, and he believed it had to be discussed for the obvious reason: "The consumption of sugar is undoubtedly increasing," wrote Allen. "It is generally recognized that diabetes is increasing, and to a considerable extent, its incidence is greatest among the races and the classes of society that consume [the] most sugar."

Allen divided the European authorities into three schools of thought on a possibly causal relationship between sugar and diabetes. Some, like the German Carl von Noorden, author of several multi-volume textbooks on diabetes and disorders of metabolism and nutrition, rejected the idea outright; some, like the German internist Bernhard Naunyn (whom Joslin had visited as a young physician to learn about the disease), thought the evidence that sugar caused diabetes was ambiguous. These physicians wouldn't blame sugar for actually causing diabetes, but did concede, wrote Allen, that "large quantities of sweet foods and the maltose of beer" favored the onset of the disease. Others, most notably the French authority Raphaël Lépine, were convinced of the causal role of sugar, and mentioned as evidence that diabetes was suspiciously common among laborers in sugar factories.

As Allen noted, however, what physicians said about sugar and diabetes and how they acted were often disconnected (as is still the case today): The majority of these authorities seemed to think that sugar had little or no role in actually causing the disease, although they were "open to accusations against sugar" when it came to the possibility that it exacerbated diabetic complications. Virtually all these physicians, however, including these same skeptical authorities, told their diabetic patients not to eat sugar, suggesting that they did, indeed, think sugar was harmful. "The practice of the medical profession is wholly affirmative" of this idea, Allen wrote. If sugar could make diabetes worse, he noted, which was implied by this near-universal restriction of sugar in the diabetic diet, then the possibility surely existed that it could cause the disease to appear in individuals who might otherwise seem healthy.

Allen's thinking had been influenced heavily by a discussion on "diabetes in the tropics" at the 1907 annual meeting of the British Medical Association. Influential British and Indian physicians working in the Indian subcontinent had discussed the high and apparently growing prevalence of diabetes among the "lazy and indolent rich" in their populations, and particularly among "Bengali gentlemen" whose "daily sustenance . . . is chiefly rice, flour, pulses, sugars."

"There is not the slightest shadow of a doubt that with the progress of civilization, of high education, and increased wealth and prosperity of the people under the British rule, the number of diabetic cases has enormously increased," observed Rai Koilas Chunder Bose, a fellow at Calcutta University, noting that perhaps one in ten of the "well-to-do class of Bengali gentleman" had the disease. Bose added that Hindu physicians had diagnosed diabetes back in the sixth century and even then had noticed the honey urine — "ants flock" around it — while observing that this was a disease "which the rich principally suffer from, and is brought on by their overindulgence in rice, flour, and sugar." Allen found this point singularly compelling. These early Hindu physicians, after all, were linking diabetes to carbohydrate consumption and sugar more than a millennium before the invention of organic chemistry and its revelations that sugar, rice, and flour were carbohydrates and that carbohydrate "in digestion is converted into the sugar which appears in the urine." "This definite incrimination of the principal carbohydrate foods," Allen wrote, "is, therefore, free from preconceived chemical ideas, and is based, if not on pure accident, on pure clinical observation."

What was unclear was whether the dietary trigger of diabetes was all carbohydrates, just refined grains (white rice and white flour among them) and sugars, sugars alone, perhaps gluttony itself, or even some other factor that predisposed the well-to-do to diabetes and protected the poor. From the discussion at the British Medical Association meeting, it was apparent that poor laborers could live on carbohydrate-rich diets without getting diabetes, whereas well-to-do Indians (and even affluent Chinese and Egyptians, as was noted by physicians at the conference) who lived on carbohydrate-rich diets easily succumbed to diabetes and seemed to be doing so at ever-increasing rates. What was the difference in their diet and lifestyle? "Unless the unknown cause of diabetes is present," wrote Allen, "a person may eat gluttonously of carbohydrate all his life and never have diabetes." Some of the physicians at the British meeting had suggested this unknown cause was the mental stress or "nervous strain" of the life of a professional — a doctor or a lawyer — compared with the relatively simple life of a laborer (as the British physician Benjamin Ward Richardson had suggested as a cause of diabetes in his 1876 book, *Diseases of Modern Life*); others suggested it was the idle life led by the wealthy and their disdain of physical activity that brought on the disease. Still others thought it was gluttony, or maybe alcohol. Sugar itself, as Allen noted, was consistently raised as a possibility.

Allen considered it likely that individuals are born with a certain innate ability to assimilate the carbohydrates in their diet and use them for energy. If the carbohydrates consumed overwhelm that ability, the excess go unused by the body and so are voided in the urine — hence the "glycosuria" or sugar in the urine that was then the principal diagnostic symptom of the disease. Maybe eating sugar somehow overtaxed this process in some people, but not all, and heavy manual labor might work to counter the effect. "If he is a poor laborer he may eat freely of starch," Allen suggested, "and dispose safely of the glucose arising from it, because of the slower process of digestion and assimilation of starch as compared with free sugar, and because of the greater efficiency of combustion in the muscles due to exercise. If he is well-to-do, sedentary, and fond of sweet food, he may, with no greater predisposition, become openly diabetic."

By the mid-1920s, the rising mortality rates from diabetes in the United States had become the fodder of newspapers and magazines; Joslin, the Metropolitan Life Insurance Company, and the New York State commissioner of health were all reporting publicly what Joslin was now calling an epidemic. When Haven Emerson, head of the department of public health at Columbia University, and his colleague Louise Larimore discussed this evidence at length at two conferences in 1924 — the American Association of Physicians and the American Medical Association annual meetings — they considered the increase in sugar consumption that paralleled the increasing prevalence of diabetes to be the prime suspect.

It wouldn't stay that way. Over the next thirty years, a series of misconceptions propagated by just a few very influential diabetes specialists, led by Joslin himself, would come to exonerate sugar almost entirely as *a* cause of diabetes, let alone *the* primary cause of the steadily increasing rates of diabetes. The argument that sugar was a cause of obesity and diabetes would be revisited again in the 1970s, but by that time the clinicians studying and treating diabetes would barely be involved.

One of the common themes in the history of medical research is that a small number of influential authorities, often only a single individual, can sway an entire field of thought. In science, young researchers are taught to challenge authority and to be skeptical of all they're taught, but this isn't the case in medicine, where the opinion of figures of authority carry undue weight. This can be particularly damaging when the state of the science is immature and the number of researchers pursuing answers is small. In the United States, Joslin became that single influential figure in diabetes, and his opinions on the subject were often treated as gospel. By the mid-1920s, Joslin had far surpassed Allen as the leading authority in the United States on diabetes, and his textbook, *The Treatment of Diabetes Mellitus,* would become the bible in the field. He published the first edition in 1916, based on what he had learned from the thousand patients he had treated at his clinic, and he and his colleagues would update it nine times by his death, at age ninety-two, in 1962.* With Joslin arguing in edition after edition of his textbook that sugar was not the cause of diabetes, the entire field would eventually accept this as truth.

By all accounts, Joslin was a remarkably dedicated physician, always working for the best interests of his patients. After insulin was discovered by researchers at the University of Toronto in 1921, Joslin's clinic pioneered its use in the United States, and he, like other physicians, quickly came to believe that insulin allowed diabetic patients to be free of the burden of severe carbohydrate restriction that until then had been thought necessary to control the disease. Perhaps more striking, juvenile diabetics, with the acute form of the disease (now known as type 1), were freed from the torturous near-starvation regimen that Allen had pioneered and upon which he had built his reputation. With insulin, both older diabetics and younger ones could eat carbohydrates, keep their blood sugar under control, and live relatively normal lives. Jos-

* The latest edition — the fourteenth, 1,224 pages long — was published in 2005.

lin's colleague Priscilla White, who specialized in treating the diabetic children at his clinic, would later say, "No child can grow up without a scoop of ice cream once a week," and insulin made this kind of indulgence possible.

Joslin recognized the value of sugar for athletes, as his colleagues at Harvard had reported about marathon runners in 1925 (to the ridicule of the *Lancet* editors).* He also recognized that consuming sugar in the form of candy, for instance, could immediately reverse the low blood sugar (hypoglycemia) or even diabetic coma that could result from poorly timed or ill-dosed injections of insulin. ("An orange is less temptation to a child than two or three pieces of sugar or even of candy," Joslin cautioned in the 1923 edition of his textbook.) Joslin believed that sugar was a valuable item in the diet, and thus unlikely to be a cause of chronic disease.

Joslin simply didn't understand that the

* In a public lecture on diabetes in 1925, according to *The New York Times,* Joslin made a point of asserting that sugar given to tired athletes renewed their vigor: "Chocolate bars for marathon runners and sugared tea for football players may result in new records," he declared.

carbohydrates in sugar had unique properties that other carbohydrates did not. He was a physician, not a nutritionist, although he had studied biochemistry for a year at Yale. He would argue that all carbohydrates were, in effect, the same — starch, grains, sugars. Joslin was the first of the many influential medical authorities who literally didn't know what they were talking about when talking about sugar; his beliefs and his ultimately successful defense of sugar in the diet would be based largely on this misconception.

As early as 1917, Joslin was using the Japanese as the singular reason to question the idea that sugar caused diabetes, and his textbook would continue to make the same argument, often in the same words, for the next forty years. "Indeed, a high percentage of carbohydrate in the diet does not appear to predispose to diabetes," he had written. "Thus, the Japanese live upon a diet consisting largely of rice and barley, yet so far as statistics show, the disease is not only less frequent but milder in that country than in this." He acknowledged that the rising death rate from diabetes in the United States coincided with rising sugar consumption, and he even had a table in the early editions of his textbook showing how sugar consumption increased step by step with diabetes mortality. "Such a marked alteration in the diet of a nation is noteworthy and deserves attention," he noted. The obvious conclusion would be to *assume* that the two "must stand in relation," he added, but the Japanese experience simply argued otherwise: "Fortunately, the dietary habits and the statistics upon diabetes of Japan would seem to save us from this error."*

Joslin came to blame the diabetes epidemic on two primary factors rather than sugar. The most obvious was obesity, because of the close association between the two conditions. Since most adult diabetics were fat, Joslin assumed that it was their fatness that made them diabetic, and he believed they got fat in the first place

* This was a natural assumption and was often made by physicians working in Asian countries as well: Isidor Snapper, for instance, who spent the World War II years in China, reported that diabetes had become a common disease among the well-to-do Chinese but was very infrequent among the poor: "It would seem that the extremely low caloric diet, consisting mainly of carbohydrates, fresh or salted vegetables and soybean flour must have had a mitigating influence upon the diabetes."

because they ate too much and moved too little. (In 1925, Joslin gave a lecture in which he blamed diabetes in part on the invention and spread of the automobile, which made people more sedentary than they had been previously and thus, he believed, fatter.)

Joslin would also come to believe that diabetes was caused by a diet rich in fat, which fed into his belief that sugar could be absolved. It was "an excess of fat, an excess of fat in the body, obesity, an excess of fat in the diet, and an excess of fat in the blood," he wrote in 1927. "With an excess of fat diabetes begins and from an excess of fat diabetics die. . . ." This was the lesson passed on as well by Cyril Long, a prominent diabetologist and dean of the Yale School of Medicine. "While there is a popular conception that an increased consumption of sugar is associated with the increasing incidence of diabetes," wrote Long, "it can be said with considerable assurance that excessive carbohydrate consumption in itself is not a direct cause of the disease." Long's view was informed by his suspicion that dietary fat was the more likely suspect.

Physicians specializing in the treatment of diabetes would come to assume that when medical textbooks used phrases like "considerable assurance," they did so based on compelling evidence, but this simply wasn't the case. Long's opinion was based almost entirely on the assertions of another profoundly influential diabetes researcher, Harold Himsworth, of University College Hospital in London, and Himsworth's assertions were based as much on his own work as Joslin's.

Like Joslin, Himsworth would have an illustrious career in medicine. In 1948, he would be named secretary of the British Medical Research Council (similar to the National Institutes of Health in the United States), a position he would hold for two decades. But he was only in his mid-twenties in 1931, when he proposed that a diet relatively rich in carbohydrates was ideal for diabetics, implying that a diet rich in fat might be a cause of the condition. "Sugar is what must be given" to treat diabetic coma, Himsworth explained, and so it stood to reason that sugar and other carbohydrates (glucose) would be valuable for any diabetic diet.

Himsworth would later report that diabetes rates had risen in Western countries in parallel with a general increase in fat consumption and a decrease in carbohydrates.* And he came to believe, as other researchers had suggested, that consuming carbohydrates helped build up an individual's ability to tolerate carbohydrate-rich foods, and that consuming the kind of fat-rich diet typically fed to diabetics did the opposite. "It would thus appear," wrote Himsworth, "that the most efficient way to reduce the incidence of diabetes mellitus amongst individuals predisposed to develop this disease would be to encourage the consumption of a diet rich in carbohydrate and to discourage them from satisfying their appetite with other types of food."

In his textbooks and articles, Joslin would describe Himsworth's "painstakingly ac-

* To make his argument that fat caused diabetes, Himsworth had to reject evidence that populations like the Inuit or the Masai, eating very-high-fat diets, also had very low diabetes rates, or at least they did at the time that Himsworth was making his claims. He did so by insisting that the evidence regarding the Masai was "so scanty" that it could be ignored, and then by misreading two articles — one on the Inuit on Baffin Island and one on the "fisherfolk" of Labrador — to claim that the Inuit, despite all evidence to the contrary, actually consumed carbohydrate-rich diets.

cumulated" data implicating fat as a cause of diabetes and so exonerating sugar. (Long described Himsworth's "very significant observations" leading to those conclusions.) Himsworth in turn would cite Joslin as the ultimate authority that sugar was not the cause of diabetes, and that fat might be. Through the 1930s and 1940s, the two constructed the scientific equivalent of a house of cards in support of their beliefs, each citing the other's observations as evidence, only to be cited in turn as the support for that evidence. Both ultimately based their conclusions largely on the incorrect assumption that sugar and other carbohydrates were equivalent in their chemical composition and thus their effect on the human body. Both returned, again and again, to the Japanese experience as the key. Here was a nation that consumed very little fat and considerable carbohydrates and had very little diabetes. Joslin took this fact as compelling evidence that carbohydrate-rich diets were beneficial; Himsworth used it to argue that fat-rich diets caused diabetes. Both exonerated sugar in the process.

Neither Himsworth nor Joslin apparently bothered to ask whether the Japanese consumed less sugar than the Americans or the British — which they did. As late as 1963, per capita sugar consumption in Japan had been roughly equivalent to the quantity consumed in England and the United States a century earlier, when diabetes was still a very rare disease in those countries as well. The Japanese experience could have been used to support the sugar/diabetes connection just as Joslin and Himsworth used it to refute the connection.

One of the many remarkable aspects of this history is that after Joslin concluded that Himsworth's fat hypothesis of diabetes was sufficiently compelling to be accepted as undisputed truth, Himsworth himself rejected it. In a 1949 lecture to the British Royal College of Physicians, Himsworth described the problem with the hypothesis as a paradox: even though populations that consumed more fat tended to have more diabetes, "the consumption of fat has no deleterious influence on sugar tolerance, and fat diets actually reduce the susceptibility of animals to diabetogenic agents." Put simply, the more fat that laboratory animals consumed to replace carbohydrates, the harder it was to make them diabetic. Now Himsworth suggested that maybe dietary fat wasn't the culprit, after all, and perhaps there were "other, more important, contingent variables" that tracked with fat in the diet. He suggested total calories as a possibility — overeating of all foods — because of the association between diabetes and obesity, and because "in the individual diet, though not necessarily in national food statistics, fat and calories tend to change together." Himsworth omitted mention of sugar, however, which is another contingent variable that tracks together with fat and calories in both national food statistics *and* individual diets.

With Joslin in the United States and Himsworth in the U.K. arguing that sugar did not cause diabetes, this statement took on the aura of undisputed truth. By the 1971 edition of Joslin's textbook, edited by his colleagues nine years after his death and now renamed *Joslin's Diabetes Mellitus,* the subject of whether or not sugar consumption caused diabetes had vanished entirely. Just as other physicians and nutritionists around the world began again to suggest that sugar was an obvious cause of obesity, diabetes, and now heart disease as well, diabetes researchers in the United States would assume *a priori* that the possibility was no longer worthy of serious attention. Rather, they would argue that obesity itself was the cause, targeting gluttony and sloth and *all* calories together, rather than sugar by itself.

Chapter 6
The Gift That Keeps on Giving

> Diabetes . . . is largely a penalty of obesity, and the greater the obesity, the more likely is Nature to enforce it. The sooner this is realized by physicians and the laity, the sooner will the advancing frequency of diabetes be checked.
>
> ELLIOTT JOSLIN, 1921

> 18 CALORIES! in a teaspoonful of sugar . . . You use up more than that getting dressed in the morning!
>
> Advertisement from Sugar Information Inc., 1962

One more lengthy digression into the science is necessary before we get back to sugar. Since the 1930s, to summarize briefly, nutritionists have embraced two ideas that ultimately shaped our judgments about what constitutes a healthy diet. These would be the pillars on which the foundation of nutritional wisdom about the impact of foods — including sugar — on obesity, diabetes, heart disease, and other chronic diseases would be based. They were both products of the state of the science of the era; they were both misconceived, and they would both do enormous damage to our understanding of the diet-disease relationship and, as a result, the public health.

The first idea was that the fat in our diets causes the chronic diseases that tend to kill us prematurely in modern Western societies. This is what Himsworth argued and Joslin came to believe about diabetes in the 1930s, and it had spread by the 1960s to researchers looking for dietary triggers of heart disease and obesity (because of the dense calories in the fat) and eventually cancer and Alzheimer's disease as well.

At its simplest, this focus on dietary fat — specifically from butter, eggs, dairy, and fatty meats — emerged from a concept that is now known as a nutrition transition: As populations become more affluent and more urban, more "Westernized" in their eating habits and lifestyle, they experience an increased prevalence of these chronic diseases. Almost invariably, the dietary changes include more fat consumed (and more meat) and fewer carbohydrates.

This isn't always the case, however, which should have been considered a critical factor in the nutritional debates that ensued. The Inuit, for instance, pastoral populations like the Masai in Kenya, or South Pacific Islanders like those on the New Zealand protectorate of Tokelau, consumed less fat (and in some cases less meat) over the course of their relevant nutrition transitions, and yet they, too, experienced more obesity, diabetes, and heart disease (and cancer as well). These populations are the counterexamples that suggest that this dietary-fat hypothesis is wrong. The same is true of populations like the French and Swiss, who eat fat-rich and even saturated-fat-rich diets but are notably long-lived and healthy. Mainstream nutrition and chronic-disease researchers would ignore these populations entirely or invoke ad hoc explanations (the French paradox, for instance) for why their experience is not relevant.

That *all* populations, without exception, consume significantly more sugar as they become affluent and more Westernized, would occasionally be considered as a competing hypothesis, as Joslin did early in his career. Until recently, though, it would typically be rejected on the basis that (1) most influential experts believed dietary fat was the problem, and (2) carbohydrates have identical effects on the human body, whether starches or sugar, and therefore on chronic-disease states, as Joslin and Himsworth believed. By this logic, populations that ate fat-poor and carbohydrate-rich diets and had low levels of obesity and diabetes (such as the Japanese) were held up as definitive evidence that fat is the problem and sugar is harmless.

The second pillar of modern nutritional wisdom is far more fundamental and ultimately has had far more influence on how the science has developed, and it still dominates thinking on the sugar issue. As such, it has also done far more damage. To the sugar industry, it has been the gift that keeps on giving, the ultimate defense against all arguments and evidence that sugar is uniquely toxic. This is the idea that we get obese or overweight because we take in more calories than we expend or excrete. By this thinking, researchers and public-health authorities think of obesity as a disorder of "energy balance," a concept that has become so ingrained in conventional thinking, so widespread, that arguments to the contrary have typically been treated as quackery, if not a willful disavowal of the laws of physics.

According to this logic of energy balance, of calories-in/calories-out, the only meaningful way in which the foods we consume have an impact on our body weight and body fat is through their energy content — calories. This is the only variable that matters. We grow fatter because we eat too much — we consume more calories than we expend — and this simple truth was, and still is, considered all that's necessary to explain obesity and its prevalence in populations. This thinking renders effectively irrelevant the radically different impact that different macronutrients — the protein, fat, and carbohydrate content of foods — have on metabolism and on the hormones and enzymes that regulate what our bodies do with these foods: whether they're burned for fuel, used to rebuild tissues and organs, or stored as fat.

By this energy-balance logic, the close association between obesity, diabetes, and heart disease implies no profound revelations to be gleaned about underlying hormonal or metabolic disturbances, but rather that obesity is driven, and diabetes and heart disease are exacerbated, by some combination of gluttony and sloth. It implies that all these diseases can be prevented, or that our likelihood of contracting them is minimized if individuals — or populations — are willing to eat in moderation and perhaps exercise more, as lean individuals are assumed to do naturally. Despite copious reasons to question this logic and, as we'll see, an entire European school of clinical research that came to consider it nonsensical, medical and nutrition authorities have tended to treat it as gospel. Obesity is caused by this caloric imbalance, and diabetes, as Joslin said nearly a century ago, is largely the penalty for obesity. Curb the *behaviors* of gluttony (Shakespeare's Falstaff was often invoked as a pedagogical example) and sloth (another deadly sin) and all these diseases will once again become exceedingly rare.

This logic also served publicly to exonerate sugar as a suspect in either obesity or diabetes. By specifying energy or caloric content as the instrument through which foods influence body weight, it implies that a calorie of sugar would be no more or less capable of causing obesity, and thus diabetes, than a calorie of broccoli or olive oil or eggs or any other food. By the 1960s, the phrase "a calorie is a calorie" had become a mantra of the nutrition-and-obesity research community, and it was invoked to make just this argument (as it still is).

The sugar industry came to embrace this thinking as the life-blood of its organization — "Which is LESS FATTENING?" a Domino Sugar advertisement asked in 1953. "3 Teaspoons of Pure Domino Sugar Contain Fewer Calories than one medium Apple." By the energy-balance logic, sugar is seen as at worst harmless and perhaps, as the sugar industry would come to argue, an ideal food for losing weight. This view was born of the assumption that obesity is caused by overeating and that all calories are the same, and the sugar industry would take full advantage. This is why it is important to understand the evolution of this thinking, how it came to be accepted as dogma, its implication, and its shortcomings.

The energy-balance idea derives ultimately from the simple observation that the obese tend to be hungrier than the lean, and to be less physically active, and that these are two deviations from normal intake and expenditure: gluttony and sloth. It was first proposed as an explanation of obesity in the early years of the twentieth century, when nutrition researchers, as we discussed, were focused on carefully quantifying with their calorimeters the energy content of foods and the energy expended in human activity. At the time, the application of the laws of thermodynamics and particularly the conservation of energy to living creatures — the demonstration that all the calories we consume will either be burned as fuel or be stored or excreted — was considered one of the triumphs of late-nineteenth-century nutrition science. Nutrition and metabolism researchers embraced calories and energy as the currency of their research. When physicians began speculating as to the cause of obesity, they naturally did the same.

The first clinician to take these revelations on thermodynamics and apply them to the very human problem of obesity was the German diabetes specialist Carl von Noorden. In 1907, he proposed that "the ingestion of a quantity of food greater than that required by the body, leads to an accumulation of fat, and to obesity, should the disproportion be continued over a considerable period."

Noorden's ideas were disseminated widely in the United States and took root primarily through the work of Louis Newburgh, a University of Michigan physician who did so based on what he believed to be a fundamental truth: "All obese persons are alike in one fundamental respect — they literally overeat." Newburgh assumed that overeating was the cause of obesity and so proceeded to blame the disorder on some combination of a "perverted appetite" (excessive energy consumption) and a "lessened outflow of energy" (insufficient expenditure). As for obese patients who remained obese in spite of this understanding, Newburgh suggested they did so because of "various human weaknesses such as overindulgence and ignorance." (Newburgh himself was exceedingly lean.) Newburgh was resolutely set against the idea that other physical faults could be involved in obesity. By 1939, his biography at the University of Michigan was already crediting him with the discovery that "the whole problem of weight lies in regulation of the inflow and outflow of calories" and for having "undermined conclusively the generally held theory that obesity is the result of some fundamental fault."

The question of a fundamental fault could not be dismissed so lightly, however. To do that required dismissing observations of German and Austrian clinical researchers who had come to conclude that obesity could only be reasonably explained by the existence of such a fault — specifically, a defect in the hormones and enzymes that served to control the flow of fat into and out of cells. Newburgh rejected this hormonal explanation, believing he had identified the cause of obesity as self-indulgence.

Gustav von Bergmann, a contemporary of Noorden's and the leading German authority on internal medicine,* criticized Noorden's ideas (and implicitly Newburgh's) as nonsensical. Positive energy balance — more energy in than out — occurred when *any* system grew, Bergmann pointed out: it accumulated mass. Positive energy balance wasn't an explanation but, rather, a description, and a tautological one at that: logically equivalent to saying that a room gets crowded because more people enter than leave.† It was a statement that described *what* happens but not *why*. It seems just as illogical, wrote Bergmann, to say children grow taller because they eat too much or exercise too little, or they remain short

* Today the highest honor of the German Society of Internal Medicine is to be awarded the Gustav von Bergmann Medal.

† In 1968, the Harvard nutritionist Jean Mayer would make the identical point with a different metaphor: "To attribute obesity to 'overeating,' " he wrote, "is as meaningful as to account for alcoholism by ascribing it to 'overdrinking.' "

because they're too physically active. "That which the body needs to grow it always finds, and that which it needs to become fat, even if it's ten times as much, the body will save for itself from the annual balance."

The question that Bergmann was implicitly asking is why excess calories were trapped in fat tissue, rather than expended as energy or used for other necessary biological purposes. Is there something about how the fat tissue is regulated or how fuel metabolism functions, he wondered, that makes it happen?

The purpose of a hypothesis in science is to offer an explanation for what we observe, and, as such, its value is determined by how much it can explain or predict. The idea that obesity is caused by the overconsumption of calories, Bergmann implied, failed to explain anything.

Obesity has a genetic basis. Identical twins, after all, are identical not just in their facial features, height, and coloring, but in body type — in the amount of fat they accumulate and where that fat goes. Body types run in families, just as hair and eye color and any other characteristics do. In 1929, the University of Vienna endocrinologist Julius Bauer confirmed the obvious when he reported that he had taken case histories from 275 obese patients and three out of every four had had at least one obese parent. (In 2004, the Rockefeller University molecular biologist Jeffrey Friedman would describe the influence of genes on obesity as "equivalent to that of height and greater than that of almost every other condition that has been studied.")

Newburgh was openly skeptical that genes could determine fat accumulation directly, let alone whether or not we're predisposed to become obese. He acknowledged that maybe "a good or poor appetite is an inherited feature," but then claimed that "a more realistic explanation" is a family tradition of serving huge portions of all-too-tasty food — "of the groaning board and the savory dish," as Newburgh phrased it. Fat parents cooked too much for their kids, and so their kids ate too much and became fat as well. Joslin, apparently, believed the same: that the children of obese parents acquired their predisposition to become obese through the eating habits passed on through the kitchen, not through their genes.

Julius Bauer, on the other hand, had spent his professional career studying and thinking about the application of genetics and endocrinology to internal medicine, a field he had pioneered with his seminal 1917 monograph, *Constitution and Disease.* He noted that this dismissive attitude demonstrated a remarkably naïve understanding of the role of genes and how genetic traits manifested themselves in living organisms. "The genes responsible for obesity," Bauer explained, must "act upon the local tendency of the adipose tissue to accumulate fat, as well as upon the endocrine glands and those nervous centers which regulate [fat accumulation] and dominate metabolic functions and the general feelings ruling the intake of food and the expenditure of energy. Only a broader conception such as this can satisfactorily explain the facts."

Bergmann, Bauer, and other European authorities wanted to know, among other things, why men and women accumulated fat differently. Even if they both eat more than they expend, why do men tend to store that fat above the waist (the beer belly) and women below? What does a caloric imbalance — Newburgh's perverted appetite — have to do with it? Why do girls put on fat as they go through puberty and in very specific places — hips and breasts — whereas boys typically lose fat and gain muscle? Why do women put on fat when they become pregnant, and, again, below the waist, not in their abdomens? (Saying the mother-to-be is eating for two — or for more than two — as would become and remain fashionable, isn't an explanation, just another observation.)

Why do women tend to gain fat during menopause or after having their ovaries removed? Endocrinologists like Bauer studying this "well known phenomenon" in animals would discuss the obvious role that female sex hormones *must* play in inhibiting fat accumulation. Newburgh ignored the animal research, while writing off the same phenomenon in a woman as caused by an inclination to indulge herself: "Probably she does not know or is but dimly aware," Newburgh wrote, "that the candies she nibbles at the bridge parties which she so enjoys now that she is rested are adding their quota to her girth."

These kinds of observations told European clinical researchers thinking about obesity in the 1920s and 1930s that hormones had to be among those critical biological factors that regulated fat accumulation and, perhaps more to the point, that caloric balance and a perverted appetite offered no meaningful explanation. "The energy conception can certainly not be applied in this realm," Erich Grafe, director of the Clinic of Medicine and Neurology at the University of Würzburg, wrote about how fat distribution differs by sex in his 1933 textbook. Double chins, fat ankles, large breasts, or even the characteristic fat deposits of the buttocks known as steatopygia in the women of some African tribes were all examples cited by Bauer and others of the local accumulation of excessive fat about which, as Grafe said, the energy conception couldn't be applied.

In a series of articles written from the late 1920s onward, Bauer took up Bergmann's thinking and argued that obesity was clearly the end result of a dysregulation of the biological factors that normally work to keep fat accumulation under check. For whatever reason, fat cells were trapping excessive calories as fat and not allowing it to escape or be used as energy by the rest of the body, if it did. And if fat cells were being driven or instructed by these biological factors to hoard excessive calories as fat, this would deprive other organs and cells of the energy they needed to thrive, leading to hunger or lethargy. These would be consequences of the fattening process, not causes. Bauer likened the fat tissue of an obese person to that of "a malignant tumor or . . . the fetus, the uterus or the breasts of a pregnant woman," all with independent agendas, and so they would take up calories of fuel from the circulation and hoard them, regardless of how much the person might be eating or exercising. With obesity, wrote Bauer, "a sort of anarchy exists, the adipose tissue lives for itself and does not fit into the precisely regulated management of the whole organism."

By 1938, Russell Wilder, the leading expert on diabetes and obesity at the Mayo Clinic and soon to become director of the Food and Nutrition Board of the National Academy of Sciences, was writing that this German-Austrian hypothesis "deserves attentive consideration," and that "the effect after meals of withdrawing from the circulation even a little more fat than usual might well account both for the delayed sense of satiety and for the frequently abnormal taste for carbohydrate encountered in obese persons. . . . A slight tendency in this direction would have a profound effect in the course of time." By 1940, the Northwestern University endocrinologist Hugo Rony, in the first academic treatise written on obesity in the United States, was asserting that the hypothesis was "more or less fully accepted" by the European authorities. Then it virtually vanished.

As the German and Austrian medical-research communities evaporated with the rise of Hitler and the devastation of the Second World War, the notion of obesity as a hormonal regulatory disorder effectively evaporated with it. The primary German textbook on endocrinology and internal medicine in the 1950s still included a discussion of this thinking, but that textbook never saw an English translation, which is significant, since the lingua franca of medical science had now shifted from German prewar to English afterward. The German-language journals from the prewar era, and with them the best scientific thinking of the era in all the disciplines relevant to both obesity and diabetes — including metabolism, endocrinology, nutrition, and genetics — would no longer be read, nor would they be referenced. In the United States, which would now dominate medical research for decades, physicians treating obese patients in their clinics and researchers studying it in the laboratory embraced the ideas of Louis Newburgh as documented facts. "The work of Newburgh showed clearly," they would say in seminars, or "Newburgh answered that" would be the response to any suggestions that obesity was caused by anything other than a perverted appetite. The postwar generation then bequeathed their belief to the generations that followed.

This perspective might have been more understandable if not for two developments. First, animal models of obesity consistently refuted Newburgh's arguments and supported the European school of thinking. The first such models were identified in the late 1930s, and they were remarkably consistent in confirming Bauer's and Bergmann's hormonal-regulatory take on obesity. These obese animals would frequently manifest what Newburgh might have described as a perverted appetite — in other words, as they grew fatter they would appear to be exceedingly hungry and consume greater amounts of food. But they would also get obese, or at least significantly fatter, even when they didn't eat more; this was true of virtually every animal model in which the researchers thought to ask what happened if the animals were not allowed to increase the amount of food they ate or eat any more food than did their lean littermates. Some of these animals would remain excessively fat even as they were being starved to death. Whatever the defect that caused these animals to accumulate fat, it obviously wasn't the result of overeating or a perverted appetite. It had to be working either to cause the fat cells to hoard calories as fat or to suppress the animals' ability to burn fat for fuel. Or maybe both.

Occasionally, researchers studying obesity — such as George Cahill, a leading authority on diabetes, metabolism, and obesity at Harvard in the 1960s — would pay attention to this research and conclude that, indeed, animals must have evolved to regulate their fat tissue carefully, and it was this system that would have to be dysregulated to create obesity. Cahill, however, felt that this was irrelevant to humans: such a regulatory system, as Cahill put it, "is also probably present in man, but markedly suppressed by his intellectual processes."

The second development, in 1960, was the development of a new technology that allowed researchers for the first time ever to measure accurately the level of hormones circulating in the bloodstream. It was the invention of Rosalyn Yalow, a medical physicist, and Solomon Berson, a physician, and was called the radioimmunoassay. When Yalow won the Nobel Prize for the work in 1977 (Berson by then was not alive to share it), the Nobel Foundation would describe it aptly as bringing about "a revolution in biological and medical research." Those interested in obesity could now finally answer the questions about which the pre-

World War II European clinicians could only speculate: which hormones were regulating the storage of fat in fat cells and its use for fuel by the rest of the body?

Answers began coming with the very first publications out of Yalow and Berson's laboratory and were swiftly confirmed by others. As it turns out, virtually all hormones work to mobilize fat from fat cells so that it can then be used for fuel. Hormones are signaling our bodies to act — flee or fight, reproduce, grow — and they also signal the fat cells to make available the fuel necessary for these actions. The one dominant exception to this fuel-mobilization signaling is insulin, the same hormone that researchers still assumed in the early 1960s to be deficient in all cases of diabetes. Insulin, Yalow and Berson reported, can be thought of as orchestrating how the body uses or "partitions" the fuel it takes in.

When blood-sugar (glucose) levels rise, the pancreas secretes insulin in response, which then signals the muscle cells to take up and burn more glucose. Insulin also signals the fat cells to take up fat and hold on to it. Only when the rising tide of blood sugar begins to ebb will insulin levels ebb as well, at which point the fat cells will release their stored fuel into the circulation (in the form of fatty acids); the cells of muscles and organs now burn this fat rather than glucose. Blood sugar is controlled within a healthy range, and fat flows in and out of fat cells as needed. The one biological factor necessary to get fat out of fat cells and have it used for fuel, as Yalow and Berson noted in 1965, is "the negative stimulus of insulin deficiency." These revelations on the various actions of insulin led Yalow and Berson to call it the most "lipogenic" hormone, meaning fat-forming. And this lipogenic signal has to be turned down, muted significantly, for the fat cells to release their stored fat and the body to use it for fuel.

A second revelation emerged in Yalow and Berson's early papers: both type 2 diabetics and the obese, they reported, tended to have elevated levels of blood sugar *and* abnormally high levels of insulin circulating in their bloodstream. Diabetes specialists like Joslin had assumed that all diabetics — whether they had the mild form (type 2) that associated with age and overweight, or the acute form (type 1) that appeared usually in children — lacked insulin, and that this was why their blood sugar could not be controlled. After all, both types of diabetes could be treated successfully, at least temporarily, with insulin therapy.

The Austrian Wilhelm Falta, a pioneer in the field of endocrinology, and later Harold Himsworth in the U.K. had reported that older, fatter diabetics seemed to be resistant to insulin's action, but diabetes specialists had paid little attention to the implications. The fact that type 2 diabetics had elevated insulin, as Yalow and Berson were now reporting, and still had high blood sugar, meant their cells must be resistant to insulin's usual blood-sugar-reducing effect. When other researchers working with Yalow and Berson's assay quickly confirmed this observation, it was clear that what we now call type 2 diabetes is not a disease of insulin deficiency (as type 1 is) — at least not at first — but of insulin resistance. It is preceded by an excess of insulin in the circulation, and that in turn may be a compensatory effect of the body's resistance to the action of that insulin.

That was just one of the critically important implications from this work. The second emerged from the observation that the obese also had high blood sugar *and* high insulin levels (what Yalow and Berson called "hyperinsulinism," though it is now more commonly known as "hyperinsulinemia"). So, if insulin is a lipogenic hormone — if it drives fat accumulation — and the obese had high levels of insulin, maybe that was why they were obese. And maybe the relationship between obesity and type 2 diabetes was not as simple as Joslin and others in diabetes research were assuming, or at least the direction of causality might be very different. Rather than obesity's causing diabetes, perhaps the same underlying physiological defect — insulin resistance and thus this hyperinsulinism — was causing both. "We generally accept that obesity predisposes to diabetes; but does not mild diabetes predispose to obesity?" as Yalow and Berson wrote in 1965 (echoing what the Portuguese physician Abel Jordão had suggested a century earlier). "Since insulin is a most lipogenic agent, chronic hyperinsulinism would favor the accumulation of body fat."

If this was true, and it certainly made sense from a biological perspective, the vital question that the medical researchers and nutritionists had to answer was: what causes insulin resistance and thus elevated levels of insulin?

It could be gluttony and sloth, as Newburgh might have argued, and it could be obesity itself, as the obesity researchers would quickly come to believe. Obesity researchers in the United States had been rejecting a hormonal hypothesis of obesity since the 1930s, if not earlier. By assuming that hyperinsulinemia and insulin resistance were caused by obesity, they could continue to believe that obesity itself is caused merely by taking in more calories than expended. This thinking left a host of problems unsolved or unexplained — insulin resistance and hyperinsulinemia, for instance, in lean individuals — but it would become widely accepted nonetheless.

Another possibility is that these elevated levels of insulin and the insulin resistance itself were caused by the carbohydrate content of our diets, and perhaps sugar in particular. Insulin is secreted in response to rising blood sugar, and rising blood sugar is a response to a carbohydrate-rich meal. That somehow this system could be dysregulated such that too much insulin was being secreted and that this was causing excessive lipogenesis — fat formation — was a simple hypothesis to explain a simple observation. And it would support an observation that had been made for millennia — that sugar was capable of providing quick energy but also inducing corpulence in those so predisposed.

These revelations led both directly and indirectly to the notion that diets restricted in carbohydrates — and restricted in sugar most of all — would be uniquely effective in slimming the obese. By the mid-1960s, these carbohydrate-restricted diets, typically high in fat, were becoming fashionable, promoted by physicians, not academics, and occasionally in the form of hugely successful diet books. Academic nutritionists led by Fred Stare and Jean Mayer of Harvard were alarmed by this and denounced these diets as dangerous fads (because of their high fat content, particularly saturated fat), suggesting that the physician-authors were trying to con the obese with the fraudulent argument that they could become lean without doing the hard work of curbing their perverted appetites. "It is a medical fact that no normal person can lose weight unless he cuts down on excess calories," *The New York Times* would explain in 1965.

This battle played out through the mid-1970s, with the academic nutritionists and obesity researchers on one side, and the physicians-turned-diet-book-authors on the other. The obesity researchers began the 1960s believing that obesity was, indeed, an eating disorder — Newburgh's "perverted appetite" — and the ongoing revolution in endocrinology, spurred by Yalow and Berson's invention of the radioimmunoassay, did little to convince them otherwise. Many of the most influential obesity researchers were psychologists, and much of their research was dedicated to studying why the obese failed to restrain their appetites sufficiently — to eat in moderation — and how to induce them to do a better job of it. The nutritionists followed along as they focused on the question of whether dietary fat caused heart disease and perhaps obesity as well, because of its dense calories. (A gram of protein or a gram of carbohydrate has four calories; a gram of fat has almost nine.) In the process, they would continue to reject any implication that sugar had fattening powers beyond its caloric content. That it might be the cause of insulin resistance — after all, *something* was — would not cross their radar screen for decades.

The sugar industry would continue to take advantage of this conventional nutritional wisdom by defending its product, as it had been doing since the 1920s, on the basis that a calorie of sugar is no more fattening or capable of causing diabetes than a calorie of any other food. As long as obesity was considered an eating disorder, this was a perfectly legitimate assumption, a gift given to the sugar industry by nutritionists and obesity researchers with the best of intentions.

In 1956, when the sugar industry embarked on a $750,000 advertising offensive to "knock down reports that sugar is fattening," they were doing so on the seemingly sound scientific basis that calories "that are spent as energy can never be deposited as fat." A photograph of President Dwight Eisenhower putting the artificial sweetener saccharin in his coffee had provoked the campaign. His doctor, as newspapers reported, had told him to avoid sugar if he wanted to remain lean. ("Sugar Bowled Over by Photo," ran the headline in *The New York Times.*) "Sugar is neither a 'reducing food' nor a 'fattening food,' " the industry advertisements responded. "There are no such things. *All* foods supply calories and there is no difference between the calories that come from sugar or steak or grapefruit or ice cream."

Almost sixty years later, when the *Times* reported in 2015 that academic researchers were doing the bidding of Coca-Cola by taking its money to fund a Global Energy Balance Network (GEBN) and "shift blame for obesity away from bad diets," this was still the argument the researchers would invoke in their defense: "Mainstream scientists understand that obesity is caused by a calorie surplus due to over-eating or under-exercising." And anyone who didn't know this was either a quack or at best held a "fringe view." Members of the GEBN were expected to be "champions of energy balance," and to "bring science to bear on the awareness for an energy balance–based solution" to the obesity epidemic. "Energy balance," the GEBN Web site noted, "is not yet fully understood, but there is strong evidence that it is easier to sustain at a moderate to high level of physical activity (maintaining an active lifestyle and eating more calories)." By implication, the problem still wasn't drinking too much Coca-Cola, or consuming too much sugar, or even consuming too much of anything; it was not being sufficiently physically active to expend those calories, a natural implication of the energy-balance thinking. For the sugar industry and the purveyors, like Coca-Cola, of sugar-rich foods and beverages, this remarkably resilient, and yet remarkably naïve, century-old conception of why some of us get fat (or are born fat) and others don't (or aren't) was, indeed, the gift that keeps on giving.

CHAPTER 7
BIG SUGAR

> If . . . every American could be induced to tip just one extra teaspoon of sugar into his breakfast coffee alone, U.S. consumption would rise 2,000,000,000 pounds annually. . . .
>
> *Forbes,* October 1, 1955

In 1928, when the sugar industry created the Sugar Institute, its first trade association, it did so not because nutritionists were attacking sugar but, rather, to address the glut of sugar that was then flooding U.S. markets. Too much sugar meant lower prices and what *The New York Times* called "cut-throat competition" among wholesalers and refiners. The mission of the Sugar Institute was, in part, to promote a new code of ethics that would get everyone in the industry working together. It would also promote directly to the public the joys and benefits of eating and drinking sugar, because getting Americans to increase their sugar consumption was a good way of bringing supply and demand in line.

Over the next three years, the Sugar Institute placed regular advertisements in newspapers and magazines, promoting sugar as a health food — a 1930s equivalent of probiotics or multiple vitamins today. In the winter and spring, Sugar Institute advertisements pitched sugar as a means to build up the immune system and fight off colds; in the summer, sugar was pitched as an enhancement of the iced beverages that keep us cool. In the fall, sugar was the solution to mid-afternoon fatigue: "Recent scientific investigations have proved that the eating of sweet cakes, a few pieces of candy, a dish of ice cream or the drinking of a sweet beverage — even a glass of water sweetened with sugar — will revive one in an amazing way."

In 1931, though, the Department of Justice sued the Sugar Institute for trying to solve the problem of cutthroat competition by using "repressive methods" to fix prices. The case went to trial in New York City, and the court ruled against the sugar industry. The sugar industry unsuccessfully appealed to the Supreme Court, which ruled that the institute had engaged in forty-five illegal practices in assuring profits for all its members. In 1936, the Sugar Institute was dissolved.

With the coming of World War II, a new crisis arose. Nutritionists had spent the last half-century coming to understand the role of vitamins and minerals in deficiency diseases — scurvy, pellagra, and beriberi, among others. This "new nutrition" research prompted a series of studies reporting that a surprising number of Americans suffered from malnutrition; their diets failed to provide them the necessary vitamins and minerals for health. In 1940, when the military draft began, 40 percent of the first million men called up for service were rejected for medical reasons, of which the primary one was extensive tooth decay. The development prompted, among other government actions, the creation of the Food and Nutrition Board of the National Research Council and its publication of the first Recommended Daily Allowances for calories, protein, and eight other nutrients, none of which, other than calories, could be found in sugar. The head of the Food and Nutrition Board, Russell Wilder of the Mayo Clinic, declared that sugar "of all foods, [was] unquestionably the worst." Two years later, when the Food and Nutrition

Board and the U.S. Department of Agriculture released the "Basic Seven" food groups — *"For Health . . . eat some food from each group . . . every day"* — sugar was still nowhere to be found.

The growing perception of sugar as "empty calories," devoid of any protein or essential vitamins and minerals, gave the government a convenient means to prepare Americans to live with the sugar rationing that would come with the war. Nutritionists and government authorities joined what the sugar industry had come to call "food faddists" in suggesting that sugar had no place in a healthy diet. One sugar industry document described these pronouncements as "sugarcoating the bitter pill of rationing," which was a clever, and apt, way of putting it. What the industry considered an attack on its livelihood — "a heavy barrage of anti-sugar propaganda" — was launched in 1942 with a government pamphlet released in preparation for rationing: It asked the question "HOW MUCH SUGAR DO YOU NEED?" and answered it unequivocally: "NONE! . . . Food experts say you really don't *need* any sugar at all."

The American Medical Association published a report by its Council on Foods and Nutrition that described sugar as a "vitamin poor" dietary constituent, which could lead to deficiency diseases by taking the place of vitamin-rich foods. The AMA council conceded that at best sugar could be harmless when consumed with nutritious foods — milk and eggs, for instance — but even then it merely " 'dilute[d] with calories' the food which is sweetened." The report concluded that "all practical means" should be "taken to limit the consumption of sugar in any form in which it fails to be combined with significant proportions of other foods of high nutritive quality." As sugar rationing kicked into effect in 1942, other authorities were even blunter about the value of sugar in the diet. "Don't complain about sugar rationing," Louis Newburgh told a reporter. "It would be a godsend if there was no sugar at all."

In their internal documents, sugar-industry executives suggested that they had simply failed to educate government officials on the "true story" of sugar. Now they had to undo the damage, before habits that would be learned during the wartime years of sugar rationing carried over into the postwar years. "Coffee without sugar today," warned one internal industry report, "in many cases will result in coffee without sugar during the post-war period."

In 1943, the industry formed a new nonprofit organization, the Sugar Research Foundation (SRF), to set the record straight.* The rationale and strategy of the SRF — "a suggested program for the cane and beet sugar industries" — were described in a document drafted by Ody Lamborn, who was president of the Coffee and Sugar Exchange of New York and would be the SRF's first executive director. "What happens when the flood-gates are opened at the close of the war?" Lamborn's document asked. "It will readily be seen that it is important not to have the mind of the American public poisoned against an invaluable and almost indispensable food — sugar."

The focus of the SRF would be educating the public on the merits of sugar, while simultaneously funding research that would "secure all known facts about sugar and its effects on and need by the human system." Members would include sugar producers, refiners, and processors, and these companies would provide the necessary funding of roughly a million dollars a year. One model

* This is the same SRF that in 1950 would discuss the spectacular success of the sugar-tobacco marriage.

for what Lamborn and the sugar industry hoped to achieve was what the California Fruit Growers' Exchange had accomplished to sell oranges and orange juice — "Who does not know of Sunkist oranges?" — and private industries such as Heinz and Campbell were achieving with their nationally branded products. The Sugar Research Foundation, befitting its name, would not indulge in any of the questionable activities that led to the demise of the Sugar Institute. Rather, it would focus on the single major challenge that the entire industry had in common — "the defense of sugar as a food and the expansion of post-war markets for sugar."

The dilemma for such an organization is one that would become common to all such industry-funded research programs and, most notably, those of the tobacco industry: how to defend and promote the use of a product — sugar, in this case — while simultaneously funding research that is ostensibly meant to secure all known facts about the product and its effect on human health. Because this research could elucidate the problematic aspects of sugar, the two goals could turn out to be mutually exclusive. Executives of the sugar industry might hope this would never happen, but there was no guarantee. If results of the research in any way challenged "the defense of sugar," the organization would have to find a way to spin its research and its program of education to make it appear as though it didn't.

By 1951, the Sugar Research Foundation, by then renamed the Sugar Association Inc. (SAI), had distributed three million dollars in research grants throughout the highest levels of academia — from Princeton and Harvard on the East Coast to the California Institute of Technology on the West. At a time when academic researchers were encouraged to work closely with industry, the SRF/SAI grants went to some of the most prominent researchers in nutrition, carbohydrate chemistry, and metabolism. The program was exceptional, and the grants themselves would regularly be written up in *Science* and other influential scientific journals. The first award went to the Massachusetts Institute of Technology (MIT): $125,000 to fund five years of research on carbohydrate metabolism. The MIT researchers would look for new industrial uses for sugar, while training a generation of young scientists in carbohydrate chemistry. MIT announced the grant along with the news that Robert Hockett, an assistant professor of chemistry, would take a leave of absence from the university to become scientific director of the SRF/SAI. The president of MIT would later say that he hoped this collaboration with the sugar industry would be a model for how industry and universities worked together in the future, and to a great extent it was.

Among the many other researchers that the sugar industry would begin supporting during the war years, two of them — Ancel Keys, at the University of Minnesota, and Fred Stare, founder of the department of nutrition at Harvard — would become lifelong friends of the industry. Stare and Keys would play critical roles in the 1960s and 1970s, defending the place of sugar in a healthy diet and arguing against the idea that it could be a cause of chronic disease.

By the early 1950s, the SAI would begin fighting public-relations battles on multiple fronts. If Americans were told that sugar caused dental caries (the technical term for tooth decay and cavities), the SAI, with the help of the researchers it was funding, would find a way to present the evidence that suggested Americans would be foolish to consume less sugar. When obesity became an issue, as it quickly did, and Americans turned to artificial sweeteners, the SAI would take on artificial sweeteners directly. The tobacco industry in the 1960s would use similar strategies to combat the public-health campaigns against smoking, and some of the players who honed their expertise on sugar — Robert Hockett most notably — would take on the same roles for the tobacco industry.*

Cavities and tooth decay had been linked to sugar directly for hundreds of years and indirectly for thousands. In the fourth century B.C., for instance, Aristotle was asking what it was about figs, a particularly sugar-rich fruit, that damages the teeth. In the sixteenth century, when sugar had become a staple of British royalty, a German traveler to London famously commented that Queen Elizabeth's teeth were black and that this was "a defect the English

* In the early 1970s, Hockett served as scientific director for the Council for Tobacco Research. In that role, he dealt with the dilemma of funding research while simultaneously promoting consumption of the product by threatening at least one investigator with a cessation of his funding if he didn't spin the interpretation of the evidence to make it less obvious that cigarette smoke was carcinogenic.

seem subject to, from their too great use of sugar." He added that the poor in England then seemed healthier than the rich, because sugar was a luxury the poor couldn't afford. Sugar "rotteth the teeth, making them look blacke, and withal, causeth many times a loathsome stinking-breath," one seventeenth-century text suggested. "And therefore let young people especially, beware how they meddle too much with it." This thinking can be found sprinkled throughout medical opinion ever after.

Still, the prevalence of dental caries remained relatively low through the mid-nineteenth century, but then it began to explode.* By the 1890s, the British Army was rejecting a "startlingly high-proportion of recruits" because of their rotten teeth. In the 1930s, researchers on both sides of the Atlantic were documenting high rates of dental caries among the poor and malnourished. "You would have to look for a long time before you saw a working-class person with good natural teeth," wrote George Orwell in *Road to Wigan Pier* in 1937. And, indeed, few had their own teeth at all after

* That the pattern was strikingly similar to that of diabetes is probably not a coincidence.

childhood. "Various people gave me their opinion that it is best to 'get shut of' your teeth as early in life as possible. 'Teeth is just a misery,' one woman said to me."

In 1939, Weston Price, a Cleveland dentist and chair of the American Dental Association's research committee, published *Nutrition and Physical Degeneration,* his seminal study of dental health around the world. As Price reported, and other researchers would confirm, isolated populations — including Swiss mountain villages, pastoral populations in Central Africa, the Inuit and First Nations people of North America, South Pacific Islanders — had nearly cavity-free teeth and retained their teeth for life, as long as they consumed their traditional diets and avoided the sugar and white flour that had come to dominate diets in the United States and Europe. "It is true that dental caries was not a major health and economic hazard until refined sugar was made available," wrote the Northwestern University chemist L. S. Fosdick in 1952. "Even today dental caries is not a major disease in those countries where refined sugar is a luxury."

The proximate cause of tooth decay had been obvious since the late nineteenth century — bacteria living in the mouth.

When sugars are present, as Fosdick put it, "they find it a nice place to live," and produce an acidic environment that eats away at the enamel of the teeth. The effect is transient and follows each meal. Hence, the more times each day we feed our bacteria, the more times each day the teeth will come under attack. The more sugar-rich or carbohydrate-rich snacks consumed during the day, the more "cariogenic" episodes. Brushing immediately after meals was known to be relatively effective at preventing cavities, but not nearly as good as avoiding sugar entirely. By the 1930s, dentists had taken to advising diets with minimal sugar as the obvious means of prevention, and one that would work even in children who may have been otherwise malnourished.

The existing science left only one significant point of controversy, which gave the sugar industry its defense. Sugar *might* not be any worse than other easily digestible carbohydrate-rich foods, particularly white flour and starches. Glucose was known to fuel the same acid-secreting bacteria as sucrose or fructose alone. Two of the very first grants given out by the SRF had gone to researchers at the University of Iowa and Harvard (Fred Stare and his colleague

Leroy Johnson) to reassess the evidence on sugar and caries formation. By 1950, the Sugar Association, Inc., was acknowledging in its internal documents that carbohydrates, including sugar, play a causal role in tooth decay, and that sugars that dissolved easily in water — sucrose and glucose — might play a bigger role than starches, though the latter point was still open to debate.

The problem, from the sugar-industry perspective, was that dentists didn't seem to care about the ambiguity and were simply telling children to avoid sugar. Hence, the "ultimate aim" of the industry's research, according to the SAI's annual report in 1950, was to "discover effective means of controlling tooth decay by methods other than restricting carbohydrate intake." Publicly, the association would argue that there was nothing unique about refined sugar, that plenty of foods would need to be restricted if prevention was the goal. If so, wrote Robert Hockett, the SAI president, then "most of the present counsel is tragically wide of the mark." An approach that would require Americans to cut down on all carbohydrates "stands little chance of success," and so it shouldn't be done. Rather, as the sugar industry was doing, more research should be funded to come up with better ways of preventing cavities on a nationwide scale — perhaps vaccines that worked against the cariogenic bacteria. In the meanwhile, the industry would argue, the only wise counsel dentists could give and should give was to recommend "prompt brushing after every meal or a simple water rinse at the earliest possible moment after taking food of any kind that will help materially in keeping down decay."

The sugar industry would adopt a similar tactic with obesity, arguing that all foods should be restricted, not just sugar — a calorie was a calorie, after all — albeit without the implication that such a tactic was sure to fail.

Whether a coincidence or not, the 1950s became the decade in which Americans started dieting en masse — or at least when the media began paying attention and low-calorie food products exploded as a food category. "Millions of Americans — male and female — were locked in the battle of the bulge," according to *Time* magazine in 1953. The American Medical Association "had described obesity as America's No. 1 health problem," noting that the thirty-four million Americans who were then overweight (according to a Gallup poll) had a higher risk of dying than the lean. By the end of the decade, *The New York Times* was reporting on "the great American dieting neurosis," while noting that one in five Americans was now "overweight" (defined as 10 percent above their "desirable" weight) and that one in three — another Gallup poll — was planning to diet, if he or she hadn't already done so (and regained, as was apparently inevitable, whatever weight had been lost).

The diet industry was now exploding, and the sugar industry perceived this as a direct threat to its viability. In 1952, some fifty thousand cases of "low-calorie" soft drinks had been sold, and sugar-free soft drinks were perceived as primarily a product to be used by diabetics. In 1959, fifteen million cases were sold; this was still a tiny percentage of the soft drink market, but the share was increasing every year.

Soft-drink manufacturers could respond — as both Coca-Cola and Pepsi quickly did — by creating their own diet soft drinks, but the sugar industry had no such option. Its only means of protecting its market share was by going on the offensive, first by defending the role of sugar in a healthy diet, even as a tool for dieting, and then by attacking the competition — artificial sweeteners — directly, as it would in the 1960s.

In 1951, the American Sugar Refining Company launched an intensive advertising campaign — the goal was nine hundred million messages, delivered in three hundred daily newspapers, Sunday supplements, and farm journals — stressing how important it was for children, in particular, to benefit from the energy contained in pure sugar. Three years later, the Sugar Association took over the effort, working through its public-relations arm, Sugar Information, Inc., which would now be dedicated to communicating the proposition that sugar was an indispensable food in *any* diet. The Sugar Association budgeted $1.8 million for a three-year advertising blitz — an "educational campaign" — and hired the legendary Leo Burnett advertising agency in Chicago to craft it.*

While physicians at Harvard, Cornell, and Stanford medical schools were now publish-

* Burnett's agency was famous, among other things, for the Jolly Green Giant, Tony the Tiger, the Pillsbury Doughboy, and the Marlboro Man. In 1998, *Time* magazine listed Burnett, the "Sultan of Sell," as among the hundred most influential people of the twentieth century.

ing in the medical journals anti-obesity diets that advocated avoiding sugar and sweets entirely, as did the occasional medical textbook, the sugar industry, reported the *Times,* was dead set on convincing the public that its product was anything but fattening. Sugar Information, Inc., with the help of Leo Burnett, would do so by taking advantage of two assumptions of the nutritionists themselves. The first, as we discussed, was that obesity was caused by the excess consumption of all calories. If so, there was nothing unique about sugar. It was "neither a 'reducing food' nor a 'fattening food,' " as the sugar-industry advertisements were now proclaiming. Assumption number two was based on the idea that hunger is a response either to low blood sugar or to the diminished utilization of glucose for fuel by the central nervous system. (The latter was an idea of Jean Mayer, working in Fred Stare's department at Harvard, and funded, at least in part, by the Sugar Association.) Both assumptions would be repeatedly refuted in experiments and would remain at best controversial for another twenty years, but nutritionists had a tendency, as they still do, to hold on to their hypotheses once adopted, regardless of the evidence that might accumulate against them. These ideas continued to suggest that foods that had the ability to raise blood sugar quickly or to be metabolized quickly — as sugar did and was — would be particularly effective at staving off hunger and thus overeating.

The sugar industry capitalized on both ideas, especially since they *seemed* logical: Because sugar contains only sixteen calories* per teaspoon (a quantity chosen by Sugar Information, Inc., perhaps because people tend to put sugar in their coffee or tea by the teaspoon), and because sugar is metabolized so quickly, it "satisfies the appetite faster than any other food. Faster even than larger portions of many other foods that supply *far more calories.*" By the industry's logic, eating sugar between meals "takes the edge off your hunger, [and] helps to overcome one of the chief causes of overweight — *overeating.*" Here's the argument as a Q&A in a Sugar Information, Inc., advertisement that ran in *The Washington Post* in 1957:

Q. How can sugar help you eat less?

A. You may remember when you were

* Sugar industry ads would occasionally say eighteen.

> small, your mother wouldn't let you have a cookie or a piece of candy before a meal because you wouldn't eat all your dinner. Perhaps mother didn't know the scientific reason, but it is a fact that *no other food stems the appetite faster than sugar. . . .* If you are trying to cut down on portions, a nibble of something sweet shortly before a meal may keep you from eating far more calories than you need at mealtime.

As an increasing proportion of the public grew overweight and then obese, and as dieting did, indeed, become a national obsession, the advertisements and their very questionable logic did the job of addressing the immediate problem confronting an industry that was dedicated to maximizing both the production and the consumption of sugar.

By the early 1960s, though, Sugar Association executives came to believe that a more direct line of attack was needed to combat the growing threat to their livelihood from the use of artificial sweeteners — particularly saccharin and cyclamates — as sugar replacements. Not only were these artificial sweeteners gaining unprecedented acceptance with weight-conscious consumers, they were also less expensive than sugar. This competitive advantage may have driven the sugar industry's response more than any other factor, leading cyclamates to be removed from the U.S. market entirely within a decade, and saccharin, if not all artificial sweeteners, perhaps irrevocably tainted as a potential carcinogen.

This particular conflict, like many with sugar, had a long history. Saccharin had been discovered in 1879, a derivative of coal tar that would be marketed as a sugar alternative, and even then an inexpensive one. Saccharin was more than five hundred times sweeter than sugar, and it could be purchased for one-tenth the cost. It had the added benefit of passing through the body without apparently being metabolized, which made it ideal for diabetics, who were told by their physicians to avoid sugar, and for the obese, who might be trying to limit calories or avoid carbohydrates. "For the first time in history," as the journalist Rich Cohen wrote about saccharin, "a food was valued not for being nutritional but for having no nutritional value whatsoever."

Then, as now, saccharin was controversial. The gist of the conflict was captured as early as 1907, when President Theodore Roosevelt had what amounted to an exceedingly short argument on its risks and benefits with Harvey Wiley, chief chemist of the Bureau of Chemistry of the U.S. Department of Agriculture. At the time, Congress had just passed the Pure Food and Drugs Act, which was the first great consumer-protection law in the United States. It had been largely motivated by Wiley's efforts to safeguard Americans from the adulteration of processed foods by dangerous chemical preservatives, and from patent medicines containing addictive and dangerous drugs. The Pure Food and Drugs Act was the founding act in a series of legislations that led to the 1930 reorganization of the USDA's Bureau of Chemistry into the Food and Drug Administration as we know it today.

Wiley believed that saccharin was unsafe for human consumption (his own research apparently failed to demonstrate otherwise) and, as he would argue to Roosevelt, that any consumer who purchased a product sweetened by saccharin had been deceived. Such a consumer "thought he was eating sugar," Wiley had said, "when in point of fact he was eating a coal tar product totally devoid of food value and extremely injurious to health." Wiley was unmoved by the argument that fruit canners, for instance, could save significant money by sweetening and preserving their products with saccharin rather than sugar. He had begun his career at the Department of Agriculture in 1883 and had been tasked then with the job of developing the domestic sugar industry. Wiley, more than any single individual, gets credit for the success of the American beet-sugar industry, having spent years of his professional life determining the optimal strains of beets to plant for different soil and climatic conditions.

Roosevelt's perspective on sugar and saccharin, however, was different. He was fat and in danger of getting fatter, and his personal physician, or so Roosevelt told Wiley, had counseled him to use saccharin daily. Hence, "anybody who says saccharin is injurious is an idiot." That was the end of the argument.

Roosevelt may or may not have been right about the long-term safety of saccharin; Wiley was certainly wrong in his contention that it was "extremely" dangerous. Roosevelt did have the better instinctive understanding of the nature of the health trade-off. For him, a nonnutritive sweetener — a "non-caloric" sweetener — seemed to be an obvious means of preventing corpulence. He correctly understood the policy question to be: which was worse, sugar or saccharin?

In 1975, when the FDA was moving toward a ban of saccharin, this is how thoughtful scientists also framed the issue. Philip Handler, head of the National Academy of Sciences, would describe it as a trade-off in his introduction to a symposium on sweeteners hosted by the NAS. As long as those who are overweight died sooner than the lean, as actuarial tables showed — "bearing out an old aphorism I learned as a graduate student," Handler said: 'The thin rats bury the fat rats' " — and assuming some weight or health benefits could be gained from consuming a noncaloric sweetener rather than sugar itself, then the question should be a risk-benefit analysis: What degree of risk from cancer or some other ailment was acceptable in the face of the benefit?

But this was not how the FDA saw it. The FDA mandate in regulating food additives focused almost exclusively on risk, as it always had. Despite Roosevelt's contention of saccharin's safety, from 1913 onward the federal government required that saccharin-containing products be plainly labeled: they could be used only "for the benefit of those to whom sugar is harmful or deleterious" or "by persons who must restrict their intake of ordinary sweets." Sugar shortages, particularly during the two world wars, would prompt increases in saccharin use as a sugar substitute, but otherwise it was marketed to and apparently used primarily by the diabetic and dyspeptic.

Cyclamates did not have saccharin's illustrious and controversial history. Sodium cyclamate was discovered in 1937 and by 1950 was being marketed in pill form by Abbott Laboratories. The compound was thirty times sweeter than sugar, as was calcium cyclamate, a sister compound, and they both lacked the bitter aftertaste that some individuals noticed with saccharin. They could also be used for cooking and baking without any loss of sweetness, which wasn't true of saccharin.

The FDA required the same labeling on products sweetened with cyclamates that it did with saccharin-sweetened products: "used only by those persons who must restrict their intake of ordinary sweets." But by the 1950s, the number of those individuals was apparently skyrocketing. Certainly the number of individuals who *wanted* to restrict their intake of ordinary sweets was. And thus was born a diet-food industry to support a nation of dieters, typically using a ten-to-one mixture of cyclamate to saccharin that would become the industry standard.

No-calorie and low-calorie soft drinks first appeared in 1952 — sweetened by cyclamate or the cyclamate-saccharin mixture. They were sold in pharmacies and groceries ostensibly for diabetics, but used widely. Coke and Pepsi released artificially sweetened diet sodas in 1963 — Tab and Patio respectively — following on the heels of Royal Crown's Diet-Rite and diet sodas from Canada Dry and Dad's Root Beer. Sales of diet sodas increased from 7.5 million cases in 1957 to fifty million in 1962, and then began doubling yearly. By 1964, they made up 15 percent of soft-drink sales, and analysts were predicting that they might someday constitute over a third of all sales.

The sugar industry responded with a million-dollar advertising campaign clearly meant to address the threat to business from diet soft drinks, claiming that artificially sweetened sodas failed to meet the nutritional needs of growing children and that "trying to lose weight by drinking them is like trying to lighten an airplane by emptying the ashtrays." (Royal Crown, which held almost 50 percent of the diet-soft-drink industry with Diet-Rite, responded with a series of ads rebutting the "sugar daddies": "If it's wrong to do millions of people a favor by taking the sugar out of cola, Diet-Rite pleads guilty.")

Publicly, the sugar industry would address the threat by looking for ways to diversify their products — continuing to fund research on the use of sugar in paints, detergents, water purification, and cigarettes, among other items — but none of these held the promise of replacing the sugar sales that were in danger of being lost to artificial sweeteners.

Privately, the industry would try to generate the evidence that the FDA needed to put the competition out of business. Although industry executives were remarkably open about this strategy, at least once it was showing signs of success. In 1969, after the Sugar Association created the International Sugar Research Foundation, John Hickson, the Foundation's vice president, described the sugar industry's position as either "find new arguments to use as leverage to force the FDA to fulfill its regulatory functions or expect to see major fractions of its markets taken over." To *The New York Times,* Hickson phrased this position in slightly more colloquial terms: "If anyone can undersell you nine cents out of 10," he said, speaking of cyclamates and saccharin, "you'd better find some brickbat you can throw at him."

That brickbat, to be precise, was a 1958 amendment to the Pure Food and Drugs Act that had been passed by Congress twenty years earlier. The original act had mandated that the FDA approve any new ingredient in processed foods as safe before it could be used, specifying that the only criterion for approval was *safety.* If a product had a safety risk, no amount of benefit from its use would work in its favor. There would be none of the trade-offs that Roosevelt had perceived or Philip Handler would later describe. A New York congressman named James Delaney chaired the congressional committee responsible for the 1958 amendment, and Delaney had recently lost a close relative to cancer. Hence, the amendment came with what would come to be called the "Delaney clause," specifying that "no additive shall be deemed to be safe if it is found to induce cancer when ingested by man or animal."

The 1958 amendment had also allowed the FDA to exempt some seven hundred existing substances from the approval process on the grounds that they were "generally recognized as safe," a designation that depended on the opinions of experts with the appropriate qualifications. These substances, which included both cyclamates and saccharin, had what would come to be known as GRAS (generally recognized as safe) status: the industry could freely use and sell them as food additives, but if new evidence came along to raise questions about their safety, the FDA would have to reassess these as well.

Between 1963 and 1969, the Sugar Association spent more than two-thirds of a million dollars (over four million today) on research designed to force the FDA to remove cyclamates from the GRAS list and have them banned. Much of the funding went to then obscure research organizations such as the Wisconsin Alumni Research Foundation (WARF) and the Worcester Foundation for Experimental Biology. The researchers at these foundations would look at the effects of saccharin or cyclamates on ingestion and excretion, metabolism, blood transport, drug interactions, the stunting of growth, cell or chromosomal damage that might lead to cancer, on sex hormones, birth defects, behavior, and even gastric distress. The aim was to find something that could prompt the FDA to reassess the GRAS status of these artificial sweeteners. If nothing else, the research reports from these institutions would keep cyclamates and saccharin in the news as a potential health hazard and increase consumer anxiety about their safety.

In May 1965, the FDA published its first review of the medical literature on cyclamates and concluded that there was little to fear. Five months later, the Sugar Association announced that WARF had published a one-page letter in the prestigious journal *Nature* suggesting that cyclamates could stunt the growth of rats — at least when the rats consumed these noncaloric sweeteners in quantities equivalent to hundreds of twelve-ounce cans of diet soda daily. This was the only study the WARF researchers would publish on cyclamates, but the two researchers involved (apparently the president and head of the biological department at WARF) continued their research through the early 1970s, first on cyclamates and then on saccharin. They reported directly to the Sugar Association and paid multiple visits to the FDA to discuss their unpublished results and why they believed that cyclamates should be banned from public use of any kind, suggesting to the FDA investigators that cyclamates were capable of causing everything from birth defects to "mental disturbance."

William Goodrich, an assistant general counsel at the FDA, would later testify to Congress that the FDA had been skeptical of the WARF research on the grounds that it had been funded by the sugar industry, which "had an understandable interest in getting cyclamates out of the soft drinks." The sugar industry lawyers, he said, had also "bombarded [him] with memoranda and scientific arguments of every sort that the product cyclamate could not generally be recognized as safe."

Finally, in 1970, researchers funded by Abbott Laboratories, at the request of the FDA, reported that high doses of cyclamate had, indeed, caused bladder cancers in male rats. The Delaney clause would now have to be invoked. A Coca-Cola executive later noted that humans would have to drink 550 cans of Fresca daily to get the equivalent dose of cyclamates as had the rats — "you'd drown before you'd get cancer," he said — but the Delaney clause did not account for whether the dosage required to cause cancer was a realistic one.

The FDA administrators had originally hoped to ban cyclamates for use in soft drinks and other foods, but to sustain their use for diabetics and obese individuals who needed to watch their calorie consumption or whose doctors suggested they avoid sugar. The pressure from food activists concerned about chemical carcinogens prevented even that compromise. (Ralph Nader's Public Citizen's Health Research Group, for instance, argued that the FDA should regard "one of its primary missions as being a cancer-prevention agency.") In October 1970, the FDA banned all use of cyclamates. Two years later, when John Hickson left the International Sugar Research Foundation to work for the Cigar Research Council, he was described in a confidential tobacco-industry memo as a "supreme scientific politician who had been successful in condemning cyclamates, on behalf of the Sugar Research [Foundation], on somewhat shaky evidence which he had been able to conjure out of Wisconsin Alumni Research Foundation."

The sugar industry almost succeeded in barring saccharin sales as well. In 1972, the FDA removed saccharin from the GRAS list, limiting its use by the food industry but allowing consumers to continue to purchase the sweetener, while the agency waited for more conclusive research. The FDA's action was based on yet another unpublished claim from the WARF researchers: that rats consuming relatively vast amounts of saccharin also developed bladder cancer.* The rats in the WARF studies, as in the cyclamate studies that had preceded it, were conceived, developed in utero, weaned, and subsequently lived their entire lives in a saccharin-rich environment, "in excess of the amount a consumer would receive from drinking 800 twelve-ounce diet sodas daily for a lifetime," *The New York Times* would explain. ("It's humanly impossible to drink 1/10th that amount in a day," said one congressman. "The first 50 cans . . . would kill you.") Chronic toxicity studies carried out in Japan, Germany, England, and the Netherlands would all show no harm from saccharin consumption, but the Delaney clause was what it was, and the FDA had its mandate.

In 1977, after Canadian researchers reported a finding similar to what the WARF researchers had claimed, the FDA moved to ban saccharin as well. It never happened, largely because the FDA succumbed to a letter-writing campaign and settled yet again for a warning label that would stay on packets of the saccharin-based Sweet 'N

* The WARF researchers did present a paper in 1974 at a symposium on sweeteners organized by the American Chemical Society.

Low, most prominently, until the year 2000. (To confuse matters, the Canadians banned saccharin but left cyclamates on the market, so Sweet 'N Low in the United States is made from saccharin and in Canada from cyclamates.)

Researchers would later realize that the physiology of laboratory rodents is sufficiently different from that of humans so that their propensity to develop bladder cancer occasionally when living on vast amounts of artificial sweeteners is not relevant to what happens to us, as the National Cancer Institute acknowledges. The FDA now considers neither cyclamates nor saccharin to be carcinogenic. In December 2000, the FDA removed the requirement that Sweet 'N Low carry a warning label, but by that time artificial sweeteners had been, indeed, irrevocably tainted. In the 1980s, when food-industry analysts were predicting a surge in diet-soda sales that failed to last, one explanation was that consumers continued to think of these substances as far more noxious than sugars and so drank sugar-sweetened beverages instead. And by then the sugar industry had successfully fought off the greatest threat to its livelihood — that it, too, could lose

GRAS status and no longer be generally recognized as safe.

CHAPTER 8
DEFENDING SUGAR*

> If we are looking for a dietary cause of some of the ills of civilization, we should look at the most significant changes in man's diet.
>
> JOHN YUDKIN, *The Lancet,* 1963

> So the real question for me as an educator is, if I go out and tell people that I think they are eating too much sugar, if I go out and tell mothers I think they should stop their kids from eating so much sugar because it is bad for them, am I going to get flak from the scientists? Or am I going

* Much of the content in this chapter about the Sugar Association and its defense of sugar was first published as an article in the November–December 2012 issue of *Mother Jones,* which I co-authored with Cristin Kearns. Cristin unearthed all the sugar industry documents on which the article and this chapter rely.

to be allowed to make that statement without travail, on the grounds that even though we do not have hard evidence to link sugar with a specific disease, we do know that a dietary pattern containing considerably less sugar, in which sugar is replaced by a complex carbohydrate, would be a much healthier diet?

JOAN GUSSOW, chairman,
Columbia University nutrition department,
1975

In 1976, John Tatem, Jr., then president of the Sugar Association, Inc., made two memorable presentations telling the story of sugar from the industry's perspective. Tatem spoke first in January to the Chicago Nutrition Association; in October, he spoke in Scottsdale, Arizona, to a meeting of the Sugar Association's board of directors.

Sugar is a healthy if not an ideal nutrient, Tatem explained at these meetings, "the purest and most economical carbohydrate available to us." In fact, as a source of inexpensive calories, sugar was a vital nutrient in the battle against famine throughout the underdeveloped world. But recently sugar had come under attack. The "enemies of sugar," Tatem said, "have charged it with contributing to every disease and physical ailment known to man, from heart disease to sweating palms."

These enemies were the "persuasive purveyors of nutritional rubbish," said Tatem, the "opportunists dedicated to exploiting the consuming public," "the promoters and quacks" who "calculatedly enlist the mass media to their ends," who "neatly apply Goebbels' 'Big Lie' technique," and who had "successfully misled a great many well-meaning advocates and media commentators." As a result of this campaign of anti-sugar propaganda, said Tatem, "sugar, once accepted almost without question, has become a highly controversial food." And if we wanted to learn the truth, we'd have to "wade through yards of pseudoscientific drivel" to do it.

Tatem wasn't fazed, or at least not publicly, by the fact that these alleged purveyors of nutritional nonsense included, among others, Walter Mertz, head of the Carbohydrate Nutrition Laboratory at the U.S. Department of Agriculture; John Yudkin, the most influential nutritionist in the United Kingdom, founder of the first dedicated department of nutrition in Europe; and the Harvard nutritionist Jean Mayer, easily the most influential nutritionist in the United States and shortly to become president of Tufts University.

Mayer had published an article in June 1976 in *The New York Times Magazine* — "The Bitter Truth About Sugar" — linking sugar not just to cavities and tooth decay but to obesity and type 2 diabetes, what Mayer called the "fat-and-forty type" of diabetes because of its association with obesity and aging. For children, Mayer suggested, sugar is quite possibly as addictive as tobacco. "The limited bill against sucrose which can be documented is sufficient to justify a drastic decrease in our consumption," Mayer had written.

At the Scottsdale meeting, four months after the *Times* had published Mayer's article, Tatem described how the Sugar Association had come to learn that *Reader's Digest* was planning to run an excerpt of it. Tatem and his colleagues had then managed to kill the excerpt, he said, first with an hour-and-a-half call to a *Reader's Digest* editor, followed by a three-page telegram to the managing editor himself. Mayer's article, according to the telegram, which was distributed to board members at the meeting, was a "scientific farce and a journalistic disgrace," and the Sugar Association could say this because "not one shred of substantiated, admissible scientific evidence exists linking sugar to the death-dealing diseases."

This was the story that the sugar industry believed, and this was the story the Sugar Association was now widely selling to the American public. "We have moved to the defensive — the defense of our primary product," Tatem said. "In confronting our critics we try never to lose sight of the fact that no confirmed scientific evidence links sugar to the death-dealing diseases. This crucial point is the lifeblood of the Association."

The war on sugar, as the newspapers would take to calling it — and in which this book is the latest offensive — had emerged fully blown in the 1960s, when the Sugar Association went on the attack to protect what Tatem later called its lifeblood. Prominent nutritionists, physicians, and laboratory researchers had begun to publish reports suggesting that sugar seemed uniquely capable of causing a cluster of metabolic abnormalities — at least in laboratory animals, if not in humans as well — that were intimately associated with both diabetes and heart disease. These reports coincided with the rise of the consumer movement and with demands from consumer activists that the Food and Drug Administration fulfill its obligations to protect the public from harmful pesticides and additives in food. In 1969, a White House Conference on Food, Nutrition and Health, convened by President Richard Nixon, called for a complete FDA review of food ingredients that were "generally recognized as safe," or GRAS substances. Sugar had been considered by the FDA — along with other "common food ingredients" such as salt, pepper, and vinegar — to be safe for any intended use. Still, like saccharin and cyclamates, it could have its "GRAS status" revoked if the FDA were given sufficient reason to worry.

The challenge to the sugar industry, as Tatem explained, was first to its credibility — "for one of the offshoots of the consumer movement has been a great weakening of public faith in the motives of business and industry" — and then to its viability. It had to respond to the charges leveled against sugar by these researchers and public health authorities, by "the enemies of sugar," as Tatem called them. "We have had to answer back to establish the facts or run the risk of being legislated out of existence."

The sugar industry won that battle in the 1970s. In doing so, it managed to shape both public opinion on the healthfulness of sugar, and how the public-health authorities and the federal government would perceive it for the next quarter century, if not, perhaps, ever since. This was one of the great public-relations triumphs of the food industry. The Sugar Association executives certainly perceived it as such.

By the mid-1980s, academic or government researchers who suggested that sugar could be a cause of heart disease or diabetes said they were risking their credibility in the process. Largely because of the sugar industry's public-relations triumph, the consumption of sugars — both sucrose and high-fructose corn syrup — did not decrease dramatically, as Jean Mayer had suggested was necessary, but, rather, saw the greatest increase in at least half a century. This was accompanied — coincidentally or not — by equally dramatic increases in the prevalence of obesity and diabetes.

What the sugar industry accomplished in the 1960s and 1970s raises vital questions about how an industry should respond when confronted with legitimate, albeit ambiguous, research suggesting that its product is dangerous. Defending your product against the dire implications of research is a natural response, as is pointing out the limitations and conflicting nature of the evidence. But does responsibility end there? Is it justified to do no more than wait and see what future research shows?

In the mid-1970s, even researchers hired as consultants by the sugar industry were telling it to do whatever experiments and clinical trials were necessary — to spend whatever money was necessary — to establish definitively whether or not sugar causes diabetes and raises the risk of heart disease. Instead, the sugar industry launched its public-relations campaign to defend sugar and attack its critics. Because this campaign succeeded, the research necessary to establish whether the dire implications were correct, or to exonerate sugar, as the case might be, was delayed for at least twenty years. It's still being done, albeit only in fits and starts. The sugar industry's campaign, however, could only succeed with the help of a nutrition-research community that had largely come to believe that dietary fat — saturated fat in particular — was the most likely cause of our chronic diseases. Understanding that development is crucial.

In the 1950s, nutrition research had turned away from its focus on the energy content and the vitamin and mineral content of foods (the "new nutrition" of the prewar years) and instead considered the possibility that certain foods could be unique causes of the chronic diseases that tend to kill us in the developed world. Heart disease was the immediate focus of this *newer* nutrition, and the growing belief that dietary fat was the cause would determine how this scientific endeavor played out. Nutritionists and other researchers — typically, cardiologists or other physicians — were making up the methods and protocols for this research as they went along. It was all new science, and very much a work in progress. In retrospect, the key players had little idea what they were doing, or how best to do it, but their conclusions shaped fifty years of nutritional dogma and still do.

Coronary disease was the focus because of the observation that more and more Americans *seemed* to be dying of heart attacks. In 1948, the American Heart Association had begun a multimillion-dollar publicity campaign to raise money for heart-disease research. In so doing, it brought to the attention of the nation what was an undeniable fact: that more Americans died of heart disease than from any other illness. This fueled the belief that the nation was in the midst of a heart-disease epidemic, and this in turn prompted nutritionists and cardiologists to wonder why. The stress of modern living was one possibility — hence, the idea that type A personalities and corporate executives were particularly susceptible — though it had nothing to do with what we eat. The cholesterol levels in our blood were another prime suspect, and it did.

Researchers had known for decades that cholesterol was a significant component of the atherosclerotic plaques that are a distinguishing feature of coronary artery disease or coronary heart disease. Russian researchers had famously demonstrated that rabbits fed high doses of cholesterol developed lesions in their arteries that looked suspiciously like atherosclerosis. (That rabbits, which are herbivores, did not naturally consume cholesterol in their diet was a fact that was occasionally raised in protest, as it should have been.) In the 1930s, Columbia University researchers created a technique for measuring cholesterol levels in the bloodstream (serum cholesterol, in the lingo) and with this analytical tool available, cholesterol became the focus of nutrition science. Researchers could easily measure the serum cholesterol of study subjects fed on different diets and see how they differed; researchers practicing the nascent science of "risk factor" epidemiology could measure serum cholesterol in thousands of individuals in large population studies — the first, famously, was in Framingham, Massachusetts — and see who later got heart disease and who didn't; physicians measured cholesterol in their patients with heart disease and compared what they saw with the cholesterol levels in their healthy patients.

By 1952, the University of Minnesota nutritionist Ancel Keys was arguing that high blood levels of cholesterol caused heart disease, and that it was the fat in our diets that drove up cholesterol levels. Keys had a conflict of interest: his research had been funded by the sugar industry — the Sugar Research Foundation and then the Sugar Association — since 1944, if not earlier, and the K-rations he had famously developed for the military during the war (the "K" is said to have stood for "Keys") were loaded with sugar. This might have naturally led him to perceive something other than sugar as the problem. We can only guess. However, it is clear that Keys was wrong about many of his conclusions, particularly regarding the role of fat and cholesterol in heart disease. Nevertheless, his thinking and the strength of his personality — both his competitors *and* his friends described him as combative and ruthless — would drive nutrition research for the next thirty years.

The American Heart Association also played a critical role in focusing on dietary fat and cholesterol as culprits, as it still does. In 1957, the AHA published a fifteen-page assessment of the evidence, compiled by some of the leading cardiologists of the era, concluding that the dietary-fat/heart-disease hypothesis was highly questionable, and castigating researchers — presumably Keys — for taking "uncompromising stands based on evidence that does not stand up under critical examination." That would be the AHA's last critical analysis. In December 1960, the organization changed its position, albeit based on no new evidence or clinical trials. An ad hoc committee, of which Keys was now a member, took the opposite position from the 1975 report, claiming instead that the "best scientific evidence of the time" suggested that heart disease was caused by the saturated fat in our diet, and that men at high risk of heart disease (overweight smokers, for instance, with high cholesterol) should eat little of it. A month later, Keys was on the cover of *Time* magazine as the face of nutrition in America, arguing that the entire country should be consuming a low-fat diet (less than half the fat we were then consuming) and that dietary fat was indisputably a cause of heart disease.

Over the next decade, researchers on both sides of the Atlantic would carry out a series of increasingly elaborate clinical trials designed to test the hypothesis that a diet that lowered our cholesterol levels would prevent heart disease and, more important, allow us to live a longer and healthier life. The results would be, at best, ambiguous. Some of the trials suggested a modest reduction in heart disease from decreasing the saturated fat content of the diet; one even suggested that it might lengthen lives. But others suggested it wouldn't, and one even suggested that eating less saturated fat would shorten our lives.* Even today, half a century later, comprehensive reviews of the connection between dietary fat and heart disease find at best "suggestive" evidence that heart-disease risk can be increased by consuming saturated fat, and often they state that the existing evidence simply fails

* This study was completed in 1973 but not officially published until 1989, because, as the lead investigator told me, "We never saw the results that we thought we would." This kind of selection bias was all too common in this research.

to support this conclusion.

Throughout the 1960s and into the 1970s, though, the media would continue the job that *Time* magazine had started, trusting the AHA to be the unbiased authority on this issue, while communicating the idea that interest in the hypothesis that saturated fat caused heart disease, and the efforts that researchers were making to test it, constituted reason enough to believe it was true. The AHA, meanwhile, would revisit its dietary-fat recommendations in a series of reports that inevitably served to support its conclusions ever more forcibly. By 1970, the AHA was advocating low-fat diets for every American, including "infants, children, adolescents, lactating and pregnant women, and older persons," despite the continued failure of the various clinical trials actually to confirm the hypothesis, or the fact that all these studies had been done in adults — particularly adult men (who are at high risk for heart disease). Women weren't studied, and so any extrapolation of the results, ambiguous as they were, to women, let alone children and infants, would be an even greater leap of faith.

Influential researchers would acknowledge in medical journals that the dietary-fat/heart-disease relationship was "an unproved hypothesis that needs much more investigation," as Thomas Dawber, a founder of the famous Framingham Heart Study, did in *The New England Journal of Medicine* in 1978. But the press, the AHA, and eventually the U.S. Congress and the U.S. Department of Agriculture treated the hypothesis as almost assuredly true, at least until definitive research came along to demonstrate otherwise.

The simplest explanation for what happened in this period was that the dietary-fat/heart-disease hypothesis had filled a vacuum, supplying a viable and seemingly reasonable answer to the question of what aspect of diet caused heart disease. Any competing hypothesis that came along after had to overcome the belief that the question had already been answered. It would have to dislodge that dogma, which was a far harder task than filling the vacuum in the first place.

Sugar entered the discussion of causation because it seemed an obvious culprit, at least to nutritionists and researchers who had not already embraced the notion that fat was to blame. The logic that sugar was likely to be causally involved was based on a series of propositions: First, that the prevalence of heart disease was increasing in Western nations (whether as dramatically as some believed or not) and increased with affluence; it was higher in developed nations than undeveloped. Second, that the same was true of the prevalence of diabetes, obesity, and hypertension (high blood pressure). Third, that these disorders are intimately related: the obese are likely to be diabetic and hypertensive and have heart attacks; those who have heart attacks are likely to be hypertensive and obese and/or diabetic; diabetics are very likely to be obese and hypertensive and very likely to die of heart attacks. So, whatever the causal factor was, it was likely to be something that accompanied affluence and was an integral part of Western diets or lifestyles, and something that could cause all these diseases, not just heart disease alone.

The dramatic increase in cigarette smoking could be responsible, for instance, and it would turn out that smoking does, indeed, raise the risk of heart disease, but it was (and still is) hard to make the argument that cigarettes cause either obesity or diabetes. Many authorities believed that cars and mechanization had made our lives less physically active, and this could be a factor as well, but it was (and is) easy to identify populations with high levels of obesity, diabetes, and hypertension that also worked very hard for a living — poor populations without the benefits of automation and mechanization.

As for diet, by far the most significant and consistent change in human diets as populations become Westernized, urbanized, or merely affluent is how much sugar they consume. Some populations also have the opportunity to consume more animal products and particularly red meat, but other populations — the Inuit, Native American tribes of the Great Plains, and African pastoralists like the Masai — were already living predominantly on animal products, and they, too, get obese, diabetic, hypertensive, and atherosclerotic as they become Westernized. All of these populations, without exception, consume significantly more sugar with this process of Westernization. (The business model of companies like Coca-Cola and PepsiCo and the sugar industry itself is devoted to making that happen.) Fat consumption may have increased in the United States since the early twentieth century, according to USDA statistics, but the reported increase was not nearly as dramatic or as certain as it had been for sugar since the 1850s. Nutritionists legitimately argued about whether the fat-consumption figures reported by the USDA — based on estimates made during the early years of World War II — were, indeed, real.

No such ambiguity existed about sugar consumption. "We now eat in two weeks the amount of sugar our ancestors of 200 years ago ate in a whole year," as the University of London nutritionist John Yudkin wrote in 1963 of the situation in England. "Sugar provides about 20 percent of our total intake of calories and nearly half of our carbohydrate." To Yudkin and others, this simple fact made sugar the prime suspect for the rising prevalence of obesity, diabetes, hypertension, and heart disease throughout developed nations.

As this argument took hold in the early 1960s, it was bolstered by observations from Israel, South Africa, and the South Pacific linking sugar intake to what appeared to be epidemic increases in diabetes prevalence — similar to what had been happening in the United States since the end of the Civil War, but much faster, over the course of a few decades.

In 1954, Elliott Joslin himself had challenged an Israeli physician, Aharon Cohen, to test Cohen's belief that genetic predisposition was not the primary cause of diabetes.

Cohen had spent the previous decade studying and treating diabetes among Native Americans in the United States and the immigrant populations that had flooded into Israel after the Second World War. These experiences had convinced him that diet played a significant role in triggering the disease in susceptible individuals. Cohen took up Joslin's challenge by comparing the prevalence of diabetes in a local immigrant population — Jews from Yemen, at the southwestern tip of the Arabian Peninsula — that had arrived in Israel in two distinct waves. The first had come in the 1930s and had been settled in Israel for a quarter-century; the second had arrived in a legendary and massive airlift known as Operation Magic Carpet that began in 1949 and brought forty-nine thousand Yemenite Jews to Israel over the course of a single year.

The Yemenites who had been in Israel since the 1930s, according to Cohen's research, had diabetes rates very similar to those of other Israelis and of populations documented in New York and elsewhere. This rate was *fifty* times higher than that of the Yemenites who had arrived in Operation Magic Carpet and had been in the country for only half a dozen years when Cohen began his research. Cohen noted that similar disparities in disease rates for hypertension and heart disease had been reported between these two waves of Yemenite immigrants. He and his colleagues then systematically queried the Yemenites about their original diets in Yemen and what they were eating in Israel, and the singular difference was not in their fat consumption. "The quantity of sugar used in the Yemen had been negligible," Cohen reported; "almost no sugar was consumed. In Israel there is a striking increase in sugar consumption, though little increase in total carbohydrates."

George Campbell, a South African physician, made a similar series of observations in two populations served by the King Edward VIII Hospital in Durban, where Campbell ran a diabetes clinic. Campbell's research was prompted by an observation he had made that was becoming increasingly common throughout Africa: The relatively affluent whites there suffered from a spectrum of chronic disease — including obesity, diabetes, heart disease, and hypertension — that was absent in rural blacks living their traditional lifestyles. This same cluster of chronic diseases, though, was becoming increasingly apparent in blacks who had moved from rural areas into towns and cities. Campbell would describe how he was "absolutely staggered by the difference in disease spectrum" between these rural and urban populations.* This difference alone seemed to rule out genetics as the primary factor in the etiology of these diseases, and suggested some aspect of diet or lifestyle was responsible.

Campbell focused his research on a population that was descended from immigrants who had arrived in the Natal region of South Africa from India in the late nineteenth century to work as indentured laborers on the sugar plantations. Four out of five of Campbell's diabetic patients, he reported, came from this Natal Indian community, many of whom were still employed in the local sugar industry. "A veritable explosion of diabetes is taking place in these

* This same comparison would be made by Campbell and others between the disease spectrum in black Africans and in blacks in the United States, who had been (forcibly) removed from Africa only a few hundred years earlier. The comparison strongly implied that something other than genetics was involved in these chronic diseases; some aspect of diet or lifestyle had to be triggering the disease that was present in the United States and relatively absent in Africa.

people," Campbell reported. He estimated that one in three middle-aged men in this population was diabetic and described this prevalence as "almost certainly the highest in the world." (As we'll see, Campbell was wrong on this account.) Although the Indian ancestry suggested a genetic predisposition among this population, Campbell noted that the prevalence of diabetes throughout India itself was only one in a hundred. So, if a predisposition existed, it had to be triggered by the local environment. Diet was again the obvious suspect. Campbell ruled out the fat content, because it was as low in this population as it was in India. He rejected the simplistic notion that these Natal Indians were merely eating too much, because the poorer members of the community were subsisting on as little as sixteen hundred calories a day — "a figure in many countries which would be regarded almost as a *starvation wage,*" said Campbell. Yet some were still "enormously fat and suffered from undoubted diabetes proven by blood tests." Once again, the amount of sugar consumed stood out: in India, the sugar consumption per capita was twelve pounds per year, compared with nearly eighty pounds for the Indians in Natal.

Campbell also compared disease rates

between the urban and rural Zulu populations, and noted that the urban Zulus were beginning to appear in his hospital with diabetes, hypertension, and heart disease, whereas these diseases were still virtually absent in the rural Zulus. The urban Zulus, Campbell reported, were eating on average ninety pounds of sugar each year; the rural Zulus consumed only forty pounds, and this number itself had increased sixfold in a decade.

Campbell's research led him to two conclusions that are worth mentioning about the appearance of diabetes epidemics in populations. First, from his study of various groups, he suggested that most could tolerate as much as seventy pounds per capita of sugar per year — roughly what Americans and the British were consuming in the 1870s — before diabetes prevalence would begin the kind of epidemic increase he was seeing among the Natal Indian and urban Zulu populations in South Africa. Second, diabetes had an incubation period similar, for example, to the time it took lung cancer to appear in cigarette smokers. From the medical histories he had taken in his clinic, Campbell noted "a remarkably constant period in years of exposure to town life" — eighteen to twenty-two years — before diabetes appeared.

By the early 1960s, the argument that sugar caused not just diabetes and heart disease but the entire cluster of chronic diseases that associated with them was being made most forcibly by two British researchers: Thomas (Peter) Cleave and John Yudkin. Whereas Yudkin was the most influential nutritionist in the U.K., if not all of Europe, Cleave was an outsider, a British naval surgeon turned director of medical research at the Institute of Naval Medicine. Cleave argued that white sugar and refined grains were equally responsible for these common chronic diseases. Yudkin focused on sugar alone. Both informed their arguments with a Darwinian perspective that was absent from discussions of the cholesterol/saturated-fat hypothesis.

Cleave had been arguing in the pages of *The Lancet* since 1940 that the more a food changes from its natural state, the more harmful it's going to be to the animal that consumes it — in this case, humans — and that sugar and refined flour were the most dramatic examples of this. In a series of articles and books, one of which was co-authored by George Campbell, Cleave invoked what he called the "Law of Adaptation," based on his reading of Darwin, to explain the epidemics of chronic disease that Campbell and others were beginning to document around the world: species require "an adequate period of time for adaptation to take place to any unnatural (i.e., new) feature in the environment, so that any danger in the feature should be assessed by how long it has been there." To Cleave, the refining of sugar and white flour and the dramatic increase in their consumption since the mid-nineteenth century were the most significant changes in human nutrition since the introduction of agriculture roughly ten thousand years before. "Such processes," he wrote about the refining of sugar and wheat, "have been in existence little more than a century for the ordinary man and from an evolutionary point of view this counts as nothing at all."

In the local populations of the kind that Campbell, Cohen, and others were studying, the changes in sugar and white-flour consumption that Americans and Europeans had experienced over a century were occurring in many cases over the span of ten to twenty years. And so their response to these foods, by Cleave's reasoning, should be that much more dramatic — higher levels of obesity and diabetes, particularly — and appearing in these exceedingly short periods of time. If researchers studied a population of African Americans or Native Americans or South Pacific Islanders, or a population of Natal Indians, as Campbell had studied, who were consuming significant amounts of sugar, and compared them with a population of European ancestry consuming the same amount, the former would exhibit a greater prevalence of obesity and diabetes because they would have had considerably less time to adapt to these foods at such relatively large levels of consumption.

Cleave believed that the refining of the sugar and flour allowed both to be easily overconsumed. Compare the teaspoonful of sugar in a single apple, Cleave suggested, with the amount of sugar commonly taken in liquid beverages. "A person can take down teaspoonfuls of sugar fast enough, whether in tea or any other vehicle, but he will soon slow up on the equivalent number of apples," Cleave wrote. "The argument can be extended to contrasting the 5 oz. of sugar consumed, on the average, per head per day [in the United Kingdom] with up to a score of average-sized apples. . . . Who would consume that quantity daily of the natural food? Or if he did, what else would he be eating?"

What's more, Cleave argued, refining increased the speed of digestion of the sugars — both sucrose *and* glucose. The pancreas in particular would be subject to an onslaught of glucose the likes of which it had never had to confront throughout human history, and Cleave believed this could easily explain the rise of diabetes over the past century. "Assume that what strains the pancreas is what strains any other piece of apparatus," wrote Cleave, "not so much the total amount of work it is called upon to do, but the rate at which it is called upon to do it. In the case of eating potatoes, for example, the conversion of the starch into sugar, and the absorption of this sugar into the blood-stream, is a slower and gentler process than the violent one that follows the eating of [any] mass of concentrated sugar."

John Yudkin was trained not only as a physician but as a biochemist as well, having earned his Ph.D. from Cambridge University with research that the French biochemist Jacques Monod would later credit as the basis of the work that led to Monod's Nobel Prize. Yudkin had developed his interest in nutrition while serving in West Africa during World War II, when he identified the cause of a skin disease among local soldiers as a vitamin deficiency. In the early 1950s, Queen Elizabeth College (shortly to become a school of the University of London) established the first dedicated nutrition program in Europe under Yudkin's leadership, and he then devoted his own research to understanding the cause and prevention of obesity and heart disease.

In 1963, in a seminal article in *The Lancet,* Yudkin took up Cleave's idea that species are adapted — "anatomically, physiologically, and biochemically" — to a particular diet and combination of foods, and that the most dramatic departures from this diet are likely to be the harmful ones. Yudkin proposed the term "diseases of civilization" to describe the cluster of diseases including obesity, diabetes, and heart disease that are common in affluent Western societies and uncommon elsewhere. (Later researchers would prefer the term "Western diseases," to avoid the implication that somehow the only civilized societies are Westernized ones.) He attributed this pattern to the relative amount of sugar consumed.

Underlying this notion, explained Yudkin in his *Lancet* article, was a series of findings coming from American biochemists and biophysicists — at the University of California, Rockefeller University in New York

City, and Yale University — implicating the carbohydrate content of the diet in heart disease, and suggesting a common pathology underlying obesity, heart disease, *and* diabetes. This research directed attention away from cholesterol as the primary factor in heart disease and the formation of atherosclerotic plaques, and focused it instead on the particles known as lipoproteins, which ferry the cholesterol around the circulation. (Today, when we talk about LDL cholesterol — the "bad cholesterol" — we are referring to the cholesterol carried around in low-density lipoproteins, LDL particles.) Cholesterol is only one of several fatlike substances that circulate in the blood. A co-traveler with cholesterol in these lipoproteins is a form of fat known as triglycerides, and different species of lipoproteins (characterized by their density) carry differing amounts of triglycerides and cholesterol.

Either of these substances could be playing a role in heart disease, as could any of the various species of lipoprotein particles themselves. Cholesterol was relatively easy to measure in the 1950s and 1960s, as this science was developing, but triglycerides were more difficult, and quantifying the lipoprotein particles required highly specialized and expensive equipment. That didn't mean that lipoprotein particles play less of a role in heart disease, only that their role was harder to determine. As Yudkin observed, research was already suggesting that they were critical actors. One way to think about this, which is how it's often discussed today, is that the lipoproteins are like buses, and the cholesterol and the triglycerides are the passengers. The question that would be hotly debated over the next thirty years, and still is to some extent, is whether it's the buses or one or another of the passengers that are doing the harm to the artery walls and therefore causing heart disease.

By the early 1960s, as the Yale and Rockefeller researchers were reporting, it was already clear that people with heart disease were more likely to have abnormally elevated triglycerides in their blood than elevated cholesterol (as measured after an overnight fast, not immediately after a meal). Another way to phrase it is that a high triglyceride count — not cholesterol — was the more common abnormality associated with heart disease. What's more, people who were likely to get heart disease but hadn't yet manifested it — those with a family history, or with diabetes (as Joslin had noted thirty years earlier), or who were merely overweight or obese — also tended to have high triglyceride levels.

All of this suggested, as Yudkin would continue to argue, that there is a pattern of metabolic and maybe hormonal disturbances, a whole cluster of them, that cause heart disease, or at least accompany it, and that that pattern of disturbances is far more profound than merely having high cholesterol. All of this suggested, as the Yale and Rockefeller research was now demonstrating, that the carbohydrate content of the diet is playing a critical role: triglycerides in the bloodstream, in particular, remain elevated when we eat carbohydrates, not fat. From this perspective, dietary fat seems to have little or nothing to do with heart disease. Yudkin considered sugar to be the obvious suspect as the carbohydrate responsible.

Over the next decade, Yudkin tested his sugar hypothesis in a series of experiments, feeding sugar or starch to laboratory animals — rats, mice, rabbits, and pigs — and reporting that sugar consumption would raise some combination of triglycerides, cholesterol, and insulin levels. He fed human subjects sugar-rich diets and reported that this raised both their cholesterol and their triglycerides, the latter more dramatically, and that it seemed to ratchet up their insulin and even make their blood cells sticky, which suggested to Yudkin that such individuals would now be more likely to have the blood clots that precipitate heart attacks.* Other researchers began studying the effect of sugar on human subjects and animals over the course of weeks to a few months; though this research continued to be suggestive, it couldn't establish whether or not sugar was truly the cause of these chronic diseases, or whether people (and the laboratory animals used in the experiments) simply ate too much of the stuff, and so got fat first and sick second.

The kind of clinical trials that were then being carried out in the United States and Europe to test the fat hypothesis were never pursued to test the sugar hypothesis.

* In the United States, Ancel Keys and his colleagues at the University of Minnesota first fed high-sugar diets to middle-aged men and also reported that their cholesterol levels rose. Keys then repeated the studies with college students and reported that the sugar-rich diets seemed benign to them, reaffirming to Keys that he was right and Yudkin was wrong. But it is possible, if not likely, that men in their forties and fifties respond differently to sugar than they would have in their late teens and early twenties.

Through the 1960s and 1970s, researchers launched ever more elaborate and expensive trials in which the subjects were randomized to diets of differing amounts or types of fat and then followed for a year or several years to see the effect: Did they have more or less heart disease or cancer? Did they live longer or tend to die prematurely? Those trials would consistently fail to confirm that eating less fat or replacing saturated fat with polyunsaturated fat could prolong lives. *No such equivalent effort would be pursued in testing sugar.* Moreover, only a few researchers were measuring the levels of circulating triglycerides in the bloodstream. Quantifying the lipoproteins in the circulation required exorbitantly expensive and arcane equipment. And so research on these "risk factors" for heart disease, as they would come to be called, was isolated to a very few laboratories.

When cardiologists and the American Heart Association thought about the role of triglycerides or lipoproteins in heart disease, perhaps not surprisingly they considered them from a physician's perspective — not what they (or we) could learn about the genesis of heart disease by studying these other substances in our blood that associate with heart disease but, rather, whether we could expect the doctors in their offices to measure them in patients. Did they have a drug they could give patients to lower elevated triglycerides, and if so, would that drug have more benefits than risks? If not, what good was it to measure triglycerides? Any physician could easily measure the cholesterol level, as could any researchers interested in studying heart disease; therefore, cholesterol is what people studied and where the AHA invested its interest.

The medical journals in England — primarily the *British Medical Journal* and *The Lancet* — published debate after debate on the role of sugar in chronic disease. ("The refining of sugar may yet prove to have been a greater tragedy for civilized man than the discovery of tobacco," one Scottish physician suggested in a letter to *The Lancet* in 1964.) Other researchers and clinicians questioned, as scientists are wont to do, the interpretation that sugar really was responsible, and discussed what studies were necessary to determine that. The American journals, like the research community in the United States, remained focused on fat and largely quiet on the sugar question.

The Sugar Association first became concerned about the emerging evidence linking sugar to heart disease and diabetes as early as 1962, but other pressing issues took precedence. The Cuban Missile Crisis, and what a Sugar Association memo refers to as the "Castro Situation," meant that financial contributions from Cuban sugar producers, until then members of the association, would no longer be forthcoming. The threat of competition from artificial sweeteners, particularly cyclamates, had made the research program on saccharine and cyclamates the Sugar Association's "top priority," the more immediate threat to the livelihood of their industry.

In 1968, when the research arm of the Sugar Association split off to become the International Sugar Research Foundation, or ISRF (and, in 1978, the World Sugar Research Organization, which is still with us today), it did so in large part, according to sugar-industry documents, to recruit more members worldwide. These would provide more financial support to combat the accumulating evidence from researchers tying sugar consumption to both diabetes and heart disease. A 1969 ISRF brochure designed to entice sugar companies to join the effort (and so pay the membership fees), titled "What's at Stake in Sugar Research," explained that the organization would focus on nutrition and public-health studies, because "misconceptions concerning the causes of tooth decay, diabetes and heart problems exist on a worldwide basis." Put simply, ISRF funds would go to combatting the notion that sugar was a unique cause of these problems. (That a certain unconditional faith in sugar is woven into the very fabric of the organization is evident today as well. The mission of the Sugar Association, as it now says on its Web site, is that of "educating health professionals, media, government officials and the public about sugar's goodness.")

The Sugar Association had plenty of help in this regard from Ancel Keys, whose laboratory had been supported by the association since the 1940s. In 1957, Yudkin had implicitly attacked Keys's work in a paper demonstrating that, among other things, sugar consumption or even the number of TVs and radios per capita tracked with heart disease in the U.K. better than the amount of dietary fat consumed. In 1970, Keys returned the favor, in a letter he first distributed widely to colleagues and then published in the obscure journal *Atherosclerosis.* He treated Yudkin as a figure of ridicule, describing his arguments as "tendentious" and his evidence that sugar rather than fat was the cause of heart disease as "flimsy indeed" and a "mountain of nonsense."

Most of Keys's criticisms were equally applicable to his own studies, which he may have known. They spoke to flaws and limitations in the research methods that the researchers themselves were just beginning to understand — the use of short-term trials to extrapolate to long-term chronic disease states, for instance, or the implication that associations between what we eat and the diseases we later get mean that the latter was *caused* by the former. But this reality didn't stop Keys from using these ideas to discredit Yudkin and his work specifically.

Ultimately, Keys built his argument against Yudkin on the first results of Keys's famous Seven Countries Study, which had just been released and went a long way to convincing nutritionists and the public that saturated fat caused heart disease (and monounsaturated fat, as in olive oil, protected against it). This was a project he had begun in 1956. Working with an international team of collaborators, Keys had compared heart-disease rates with diet in sixteen populations in Italy, Yugoslavia, Greece, Finland, the Netherlands, Japan, and the United States. Ironically, Keys's study was the first one ever that made an attempt to measure directly both sugar and fat consumption in different populations. The conclusion was that, of all the various dietary factors measured in these populations, the two that tracked best with heart disease — as Yudkin might have predicted — were sugar and saturated fat. These are two macronutrients, along with animal protein, that populations tend to (but don't always) consume in greater quantity as they become Westernized and more affluent. Because the association that emerged from the Seven Countries Study seemed to be *slightly* stronger for saturated fat than for sugar, and because populations in the study that ate a lot of one tended also to eat a lot of the other, Keys now suggested that this was "adequate to explain the observed relationship between sucrose and [coronary heart disease] without recourse to the idea that sucrose was somehow involved in the etiology" — i.e., that sugar caused it. This was speculation, by any account, but Keys made it nonetheless. "None of what is said here should be taken to mean approval of the common high level of sucrose in many diets," he said in his takedown of Yudkin, yet he insisted that his rival "has no theoretical basis or experimental evidence" to support his claims.

Four years later, when Keys and his wife, Margaret, co-authored a diet book based on their belief in the healing powers of Mediterranean eating patterns, they insisted that Yudkin was "alone in his contentions," at least among academic researchers, and added, "Yudkin and his commercial backers are not deterred by the facts; they continue to sing the same discredited tune."

It's hard to overemphasize how the existence of the dietary-fat hypothesis influenced thinking on the sugar hypothesis and the evolution of the controversy. Researchers typically assumed that if Keys was right, Yudkin was wrong, and vice versa. (The scientific conflict wasn't helped by the fact that "there was quite a bit of loathing" personally between Yudkin and Keys, as one of Yudkin's colleagues would later phrase it.) Critical pieces of evidence would be viewed from one perspective only, and usually that of supporters of the saturated-fat hypothesis. During the Korean War, for instance, pathologists doing autopsies on American soldiers killed in battle noticed that many had significant plaque buildup in their arteries, even though they were only teenagers. The Koreans killed in battle did not. This was later attributed to the fact that the American soldiers ate plenty of butter, meat, and dairy products — all rich in saturated fat — and the Korean soldiers did not. But disparities in sugar consumption could also, obviously, have explained what was seen (as, of course, could other factors as well): as late as the 1950s, per capita sugar consumption in Korea would have been as low as or probably lower than sugar consumption in the United States a century earlier.

When researchers realized that the French had relatively low rates of heart disease despite a diet that was rich in saturated fats, they wrote it off as an inexplicable "paradox," and ignored the fact that the French traditionally consumed far less sugar than did populations — the Americans and British, most notably — in which coronary disease seemed to be a scourge. At the end of the eighteenth century, French per capita sugar consumption was less than a fifth of what it was in England. At the end of the nineteenth century, even after the beet-sugar revolution, France was still lagging far behind both the British and the Americans — thirty-three pounds for the French compared with eighty-eight for the English and sixty-six for Americans. ("Sweetness does not seem ever to have been enshrined as a taste to be contrasted with all others in the French taste spectrum — bitter, sour, salt, hot — as it has in England and America," wrote Sidney Mintz. "It is not necessarily a mischievous question to ask whether sugar damaged English cooking, or whether English cooking in the seventeenth century had more *need* of sugar than the French.")

Journalists would write about the potential evils of sugar, but then write off the idea that it could cause heart disease — as the *New York Times* personal-health reporter Jane Brody did, for instance, in a 1977 article entitled "Sugar: Villain in Disguise?" — on the basis that the notion "does not have widespread support among experts in the field, who say that fats and cholesterol are the more likely culprits."

Whereas American researchers and observers tended to side with Keys and his dietary-fat hypothesis, Europeans were more open-minded. "Although there is strong evidence that dietary fats, particularly the saturated ones, play an important role in the etiology of [coronary heart disease], there is no proof that they are the only or the main culprit," wrote Robert Masironi, a heart-disease researcher at the World Health Organization and later president of the

European Medical Association. "As regards the relationship of sugars to cardiovascular diseases, it must be borne in mind that these nutrients have common metabolic pathways with fats. Disturbances in carbohydrate metabolism may be responsible for abnormal fat metabolism and may therefore act as a causative factor in the development of atherosclerosis and of coronary disease."

In 1971, Yudkin retired from his position as chair of the nutrition department at the University of London, hoping to devote his time to research and writing. The university administrators replaced him with the South African nutritionist Stewart Truswell, who believed and argued publicly that Keys's dietary-fat hypothesis was assuredly correct and that people should change their diets accordingly. Under Truswell's leadership, the department broke its agreement to give Yudkin an office and allow him to keep his laboratory, and that ended his research career. Yudkin instead spent the first year of his retirement writing a popular polemic against sugar that was published in 1972 as *Pure, White and Deadly* in England and *Sweet and Dangerous* in the United States.

While Yudkin's work failed to move the medical-research community in the United States to embrace either him or his sugar theory, publication of his book was reported by the media: "Sugar — The Question Is, Do We Need It at All," read the *Times* headline. The press attention in turn prompted the U.S. Senate to get involved. In April 1973, a Senate subcommittee headed by George McGovern (and advised by Jean Mayer) held a congressional hearing on sugar in the diet, diabetes, and heart disease.

The testimony came from an international panel of researchers. Yudkin testified, as did Aharon Cohen, George Campbell, Peter Cleave, and Peter Bennett, a National Institutes of Health diabetes researcher working with the Pima population of Native Americans in Arizona. Bennett testified that the Pima had perhaps the highest rates of diabetes of any population ever studied. "The only question that I would have," Bennett said, "is whether we can implicate sugar specifically or whether the important factor is not calories in general, which in fact turns out to be really excessive amounts of carbohydrates." Walter Mertz, head of the Carbohydrate Nutrition Laboratory at the U.S. Department of Agriculture, also testified, as did his colleague Carol Berdanier, explaining that refined sugar seemed to play particular havoc with health, at least in laboratory rats. It elevates blood sugar and triglycerides specifically, and causes them to become diabetic, Berdanier told the congressmen, "and they die at a very early age."

The International Sugar Research Foundation responded the following March by hosting a conference in Washington, D.C. — "Is the Risk of Becoming Diabetic Affected by Sugar Consumption?" — and inviting to speak only researchers who were outwardly skeptical of the sugar–diabetes–heart disease connection. Absent from the list, therefore, were any of the researchers who had testified at McGovern's hearings and would have argued that the evidence was compelling. (The rationale: "The research and findings of these scientists are well known to the ISRF staff and members of the Foundation.")

Even the researchers recruited to speak at the conference, skeptical as they were of the sugar hypothesis, agreed that some significant percentage of individuals might be particularly sugar-sensitive, and these would experience an increase in heart-disease risk unless they restricted their sugar consumption. "From the dietary point of view," said the Belgian nutritional chemist Jean Christophe, one of the speakers, "the fact that sucrose increases serum triglycerides in some patients . . . could make imperative its restriction." A review of the conference published in a diabetes journal, which the ISRF shared with its members, concluded, "All those present agreed that a large amount of research is still necessary before a firm conclusion can be arrived at, and various suggestions were made about future research."

In September 1975, the International Sugar Research Foundation reconvened in Montreal to discuss research priorities with scientist consultants hired to point them in the right direction. It was clear now that the industry was in trouble. As John Tatem of the Sugar Association reported at the meeting, the amount of sugar sold by the industry in the United States and thus apparently consumed had dropped by 12 percent in the previous two years alone (from 102 pounds per capita to ninety), and a major factor was "the impact of consumer advocates who link sugar consumption with certain diseases."

After the Montreal conference, the ISRF disseminated a memo to its members focusing on the recommendations of Errol Marliss, a University of Toronto diabetes specialist, implying that these would be embraced by the foundation. "It is in the best interests of the industry to establish definitively what contribution sucrose can and does make to the course of diabetes — and other diseases — to place it in context," Marliss had said and the ISRF reported. "This will require the support of well-designed research programs. Such research programs *might* produce an answer that sucrose is bad in certain individuals, and if well designed, may allow for the recommendation of specific amounts to those individuals. . . . The foregoing could well be expensive in terms of the research investment, and should be undertaken in a sufficiently comprehensive way as to produce results. A gesture rather than full support is unlikely to produce the sought-after answers."

A gesture is all the sugar industry would offer. By 1975, U.S. sugar companies were pulling their support from the ISRF, disagreeing on how research money should be spent. Instead of pooling funds at an international level — "the effort to unite the world for sugar research has been a dismal failure," as Tatem reported to his board of directors — the Sugar Association would now take back control of research in the United States and get the money to do so from local sugar-using industries — eventually enlisting, among others, Coca-Cola, Hershey, General Foods, General Mills, Nabisco, Life Savers, Quaker Oats, M&Ms/Mars, PepsiCo, and Dr Pepper.

First, though, the Sugar Association hired the legendary Madison Avenue public relations firm Carl Byoir and Associates to design a public-health campaign that would "establish with the broadest possible audience — virtually everyone is a consumer — the safety of sugar as a food." (The PR firm and the Sugar Association submitted an application to the Public Relations Society of America for its 1976 Silver Anvil Award, the most prestigious honor in the PR industry, awarded for "the forging of public opinion," and Byoir's sugar-defense campaign would win it.) Point one was the recruitment of a Food and Nutrition Advisory Committee (FNAC) that would be composed of well-respected authorities in medicine, nutrition, and dentistry, all apparently willing to defend sugar as necessary to the public. To John Tatem and the sugar industry, they were "eminent and objective medical scientists."

Working to the sugar industry's advantage, once again, was the rising support for the belief that saturated-fat consumption and elevated levels of serum cholesterol were the likely causes of heart disease. At a time when Henry Blackburn, a colleague of Ancel Keys at Minnesota, was writing in *The New England Journal of Medicine* that "two strikingly polar attitudes persist" on the subject of diet and heart disease, "with much talk from each and little listening between," and when the National Institutes of Health had just launched two massive, unprecedented clinical trials, at a cost of more than a quarter-billion dollars, to test, albeit only indirectly, the dietary-fat/cholesterol hypothesis, the Sugar Association and the ISRF would build their scientific defense against sugar on the belief that saturated fat had already been proved to be the causative agent of heart disease. (Tatem would even suggest in a letter to the editor of *The New York Times,* never published, that some "sugar critics" were motivated merely by wanting "to keep the heat off saturated fats.")

When the Sugar Association needed an authority on heart disease for the FNAC, it enlisted Francisco Grande, who worked closely with Keys at the University of Minnesota. Keys and Grande had co-authored over thirty papers together, most of them either supporting the presumed relationship between dietary fat and heart disease or trying to explain away the evidence implicating sugar. A second heart-disease authority on the FNAC was the University of Oregon nutritionist William Connor, the leading proponent of the idea that dietary cholesterol caused heart disease.

For a diabetes expert, the FNAC recruited Edwin Bierman of the University of Washington. Bierman had been almost single-handedly responsible for convincing the American Diabetes Association to liberalize the amount of carbohydrates recommended in diabetic diets and to effectively ignore the sugar content. Bierman also professed an apparently unconditional faith that it was high cholesterol levels that caused heart disease, and this implicated the saturated fat in our diets, not sugar.

Bierman's role, both for the Sugar Association and working on his own, was absolutely pivotal in assuring that little research effort was expended on the possible causative role of sugar in diabetes. Bierman was unequivocal in his belief that sugar and other carbohydrates played no role in the development of diabetes, other than perhaps providing excess calories. He shaped the American Diabetes Association's nutrition guidelines, taking the ADA's focus away from sugar, when the ADA was (and still is) involved in setting the diabetes research agenda through its own funding and the significant advocacy/advisory role it plays. He also rejected the idea that sugar had any significant role in causing diabetes when he co-authored, with the epidemiologist Kelly West, a section on obesity and nutritional factors in a 1976 report by the National Commission on Diabetes — *The Long Range Plan to Combat Diabetes* — that has influenced the federal government's diabetes research agenda ever since. Some researchers, Bierman and West acknowledged, had "argued eloquently" that refined carbohydrates such as sugar could be a precipitating factor in diabetes (citing Peter Cleave and Aharon Cohen, but not Yudkin). They did not find the idea compelling, however, and omitted any further study of the role of sugar from their research recommendations. "A review of all laboratory and epidemiologic evidence," they wrote, "suggests that the most important dietary factor increasing the risk of diabetes is total calorie intake, irrespective of source." In an equally influential 1979 review published in *The American Journal of Clinical Nutrition,* Bierman would insist, "There is no known biological basis for the hypothesis that would relate higher sucrose or carbohydrate intakes to the causation of diabetes."

The point man for the Sugar Association's Food and Nutrition Committee was Fred Stare, founder and longtime chairman of the department of nutrition at the Harvard School of Public Health. The sugar industry had been supporting Stare and his department since the early 1940s, and the International Sugar Research Foundation estimated that its grants to Stare (to study the relationship between blood sugar, appetite, and obesity) had resulted in the publication of thirty research articles and reviews between 1952 and 1956 alone. In 1960, when Stare's nutrition department broke ground on a new five-million-dollar building, it was paid for largely by private donations, including the "lead gift," as Stare described it, of $1.026 million from the General Foods Corporation, the maker of Kool-Aid and the Tang breakfast drink.

By the late 1960s, Stare had become, in academia, the most public defender of sugar — it was not even "remotely true," he would write, "that modern sugar consumption contributes to poor health" — while his department received funding from the sugar industry, the National Confectioners Association, Coca-Cola, PepsiCo, and the National Soft Drink Association. (Tobacco-industry documents reveal that Stare's department, at his request, also received money from the Tobacco Research Council, specifically to fund projects that might exonerate cigarettes as a cause of heart disease.) Stare freely acknowledged that he did not use sugar in his coffee or cereal; he was saving the calories, he said, for a martini at night. But he also argued that it was unsound "and may be hazardous" to recommend that anyone, including children, avoid sugar, on the grounds that if they did they would be likely to replace it with saturated fat, "and that, I hope, everyone will agree, is not desirable."

The Sugar Association repeatedly turned to Stare and his Harvard credentials to counter any anti-sugar sentiments in the press — "plac[ing] Dr. Stare on the AM America Show," as internal memos reveal, and "do[ing] a 3 1/2 minute interview with Dr. Stare for 200 radio stations." In using Stare as its front man to dismiss anti-sugar sentiments publicly, the Sugar Association noted, it was "able to keep the sugar industry in the background" and so keep Stare's conflicts of interest in the background as well.

Ultimately, the FNAC members would be most useful as authors of an eighty-eight-page white paper, "Sugar in the Diet of Man," a compilation of the evidence and arguments going back into the 1930s that could be used to counter the research put forth by Yudkin, Mayer, Cohen, Campbell, Cleave, and the other "enemies of sugar." Stare wrote the introduction and edited the document. Grande wrote the chapter on heart disease, exonerating sugar as a cause. Bierman co-wrote the chapter on diabetes with Ralph Nelson of the Mayo Clinic, doing the same. "The causes of primary diabetes mellitus in man remains [*sic*] unknown," Bierman and Nelson wrote, but "there is no evidence that excessive consumption of sugar causes diabetes." (What made this position on sugar typically perplexing is that Bierman and Nelson didn't actually believe that diabetics should eat sugar, because it was bad for them, a point that they made in two short sentences in the eight-page chapter: "Simple sugars should still be avoided," they wrote, and sucrose is very much a simple sugar.)

The Sugar Association eventually disseminated at least twenty-five thousand copies of "Sugar in the Diet of Man." When newspaper food editors met for a conference in Chicago in 1975, copies of the white paper were included in their press packets.

(The sugar industry hosted a session there that included a talk by Phil White, a former student of Fred Stare's, who was then working as director of the department of foods and nutrition at the American Medical Association. John Tatem, who hosted the session, insisted that the subject of discussion was not sugar per se but rather food faddism in general and the many commodities, of which sugar happened to be just one, that were "falsely maligned by this element of pseudo-scientists.") When the report was sent to the press, it was accompanied by a lay summary written by a health journalist and a press release with the headline "Scientists Dispel Sugar Fears."

As with Stare's placement on radio and TV shows, the Sugar Association's role in preparing and funding the document were kept well in the background. Sugar Association documents suggest that the FNAC activities and the report itself were funded entirely by the sugar industry, at significant cost, but no such acknowledgment appeared on the document. A confidential memo to "hold and use for inquiries" about bias or conflict of interest in the report was sent by the Sugar Association to directors of communications at sugar companies across the country. According to the memo, Stare had come up with the idea for the white paper and asked the SAI to fund it, so they paid for his research time "as we would with any research project" and "purchased reprints," the twenty-five thousand copies distributed.

In November 1976, Stare's copious conflicts of interest were finally exposed in an article by Michael Jacobson, founder of the Center for Science in the Public Interest, and two colleagues, entitled "Professors on the Take." "In the three years after Stare told a Congressional hearing on the nutritional value of cereals that 'breakfast cereals are good foods,' " Jacobson and his colleagues wrote, "the Harvard School of Public Health received about $200,000 from Kellogg, Nabisco, and their related corporate foundations." ("A lot of the public, and unfortunately some of my colleagues, think I'm a monster," Stare would later acknowledge, "a paid tool of the food industry.") By 1976, however, Stare was no longer necessary for the public-relations campaign, and the Sugar Association could turn to an FDA document that took up where "Sugar in the Diet of Man" left off.

While Stare and his colleagues were drafting "Sugar in the Diet of Man," the FDA would launch its first review of whether sugar could be considered "generally recognized as safe" (GRAS). These GRAS reviews, requested by the White House after President Nixon's 1969 Conference on Food, Nutrition and Health, had been subcontracted by the FDA in 1972 to the Federation of American Societies of Experimental Biology, which in turn had created a committee of eleven members — the Select Committee on GRAS Substances (SCOGS) — to vet hundreds of food additives, from acacia to zinc sulfate. Over the course of five years, SCOGS would submit seventy-two "comprehensive reports" to the FDA, covering 230 substances that the FDA had been given reason to believe might not be as safe as thought.

This committee would officially review the science, pro and con, on sugar. Despite a stated sensitivity to industry influence in the process ("Avoidance of even an appearance of conflict of interest was emphasized," the SCOGS members would later write), the chair of SCOGS, and thus of the committee reviewing sugar for the FDA, was George W. Irving, Jr. Irving was a biochemist and a longtime member and chairman (for two years beginning in 1969) of the scientific advisory board of the International Sugar Research Foundation. Another member of SCOGS, Samuel Fomen, a University of Iowa professor of pediatrics, had received sugar-industry funding to study the role of sugar in infant feeding from 1970 to 1973.

According to the FDA guidelines, the committee could pronounce a substance to be hazardous — not generally recognized as safe — if it found "credible evidence of, or reasonable grounds to suspect, adverse biological effects . . . in whatever information was available." The committee members apparently decided, however, that if a subject was sufficiently sensitive, as sugar was ("If sucrose was to be declared a health hazard," they would later write, "what should be done about glucose, fructose, honey?"), they could decide that ambivalent evidence was reason enough to decide against the potential health-hazard conclusion.

Whether we consider this right or wrong, ethical or unethical, the committee's review of sugar relied heavily on the Sugar Association's "Sugar in the Diet of Man" and its authors. In January 1976, the Sugar Association obtained a copy of the "tentative conclusions" of the SCOGS committee, which was then disseminated to the members of FNAC with an "urgent request to review" and the anticipation that Stare and his colleagues would "identify pertinent missing and faulty data as well as possible misinterpretation of background information." But even the tentative conclusions were sugar-industry friendly. The section on sugar and heart disease said "conflicting results" were found, and cited fourteen such studies, one of which was Francisco Grande's chapter in "Sugar in the Diet of Man"; five either came from Grande's lab itself or were sugar-industry-funded studies. The single paragraph on diabetes in the SCOGS review acknowledged that studies "suggest that long term consumption of sucrose can result in a functional change in the capacity to metabolize carbohydrates and thus lead to diabetes mellitus," but then said that "recent reports tend to contradict" this. Of the four contradictory reports cited, one was Ed Bierman's chapter with Ralph Nelson in "Sugar in the Diet of Man," and two others were studies from Bierman's laboratory.

The revised version of the SCOGS review, released a year later, concluded that reasonable evidence existed to conclude that sugar caused tooth decay, but not that it was a "hazard to the public" in any other way, at least not at the levels then being consumed. It described the evidence linking sugar to diabetes as "circumstantial," and said there was "no plausible evidence" that it was related to the disease, other than as a source of excess calories. The report described the evidence linking sugar to cardiovascular disease as "less than clear." "Furthermore," it explained, "it would appear that the primary dietary factors involved in cardiovascular disease are the nature and amount of fat in the diet. Thus, the role of sucrose in cardiovascular disease appears to be secondary although it may represent a potentiating factor in its etiology."

The one cautionary note in the SCOGS review, other than the link to cavities, was that the use of sugar in the food and beverage industries had been increasing, and that, should these trends continue, all bets were off: "It is not possible to determine without additional data whether an increase in sugar consumption . . . would constitute a dietary hazard."

The SCOGS reviewers then thanked the Sugar Association for its help in "contribut-[ing] information and data" to the report, prompting John Tatem to remark later that, though he was "proud of the credit line, I think we would probably be better off without it." The report itself was signed by Irving, the former chairman of the ISRF's scientific advisory board.

Before releasing the report in January 1977, the FDA held a public hearing to discuss it. Sheldon Reiser, director of the USDA's Carbohydrate Nutrition Laboratory, and his colleagues submitted what they considered "abundant evidence" showing that "sucrose is one of the dietary factors responsible for obesity, diabetes, and heart disease." As they would later explain in a letter to *The American Journal of Clinical Nutrition,* clearly some portion of the American public could not tolerate a diet high in sugar and other carbohydrates — perhaps fifteen million adults at the time, they estimated. This alone, they had argued to the SCOGS panel, was reason to restrict sugar consumption by "a minimum of 60 percent" and urge that "a national campaign be launched to inform the populace of the hazards of excessive sugar consumption."

The members of the SCOGS panel, however, stood by their conclusions, despite "loudly proclaim[ing] the imperfectability" of expert committees like their own. They had done the "best [they] could," they later wrote, "under an enormous number of uncertainties and constraints."*

* These constraints included the limited amount

The Sugar Association, on the other hand, would pronounce the FDA effort definitive and tout the SCOGS report as a combination of salvation and exoneration. The SCOGS report had described the evidence against sugar variously as ambiguous, less than clear, or circumstantial, but the Sugar Association translated those caveats as synonymous with "nonexistent." Tatem distributed a memo to the members of the association, suggesting that the SCOGS report "should be memorized" by the staff of any company associated with the sugar industry. "In the long run," he said, "the GRAS report cannot be sidetracked, and you may be sure we will push its exposure to all corners of the country."*

"Sugar is Safe!" proclaimed a Sugar As-

---

of research, the "limitations of experimental designs," "the tangled web of social consequences associated with the introduction or withdrawal of a commercially added food ingredient," and "the continuous progression of scientific theories and empirical findings."

* In May 1976, when the Public Relations Society of America awarded its Silver Anvil Award to the Sugar Association and Byoir and Associates for the advertising campaign in defense of sugar, the society emphasized the campaign's "ability to stem

sociation advertisement about the FDA report. "Sugar does not cause death-dealing diseases. . . . There is no substantiated scientific evidence indicating that sugar causes diabetes, heart disease or any other malady." The ad ended with a caution to the unwary consumer: "The next time you hear a promoter attacking sugar, beware the ripoff. Remember he can't substantiate his charges. Ask yourself what he's promoting or what he is seeking to cover up. If you get a chance, ask him about the GRAS Review Report. Odds are you won't get an answer. Nothing stings a nutritional liar like scientific facts."

The Sugar Association did get around to funding research on diabetes, but it was nothing like the concerted effort that the scientist-consultants had argued for prior to publication of the SCOGS report. Between 1976 and 1978, the sugar industry — via the Sugar Association and the ISRF — budgeted sixty thousand dollars each year

---

the flow of reckless commentary" about sugar, and singled out the conclusions of the SCOGS report as an accomplishment that would make it "unlikely that sugar will be subject to legislative restriction in coming years."

to paying Fred Stare and his fellow Food and Nutrition Advisory Committee members, and between 1975 and 1980 it spent $655,000 on more than a dozen research projects, designed, as the industry documents put it, to "maintain research as a main prop of the industry's defense." These research proposals had to be vetted first by the FNAC members, and then by commissions that included members of the sugar industry itself and of companies such as Coca-Cola and Hershey that constituted "contributing research members." Perhaps not surprisingly, virtually all the money went to proposals that set out to exonerate sugar and to sugar-friendly researchers or simply friends of the FNAC members. (One study, at the Massachusetts Institute of Technology, proposed to explore whether sugar could be shown to boost serotonin levels in the brains of rats, and thus "prove of therapeutic value, as in the relief of depression.")

Two researchers who received Sugar Association money for their work during this period — Ron Arky of Harvard, a friend and medical-school classmate of Bierman's, and Paul Robertson, a student of Bierman's at the University of Washington — both described the research philosophy of the

Sugar Association in later interviews as a token gesture. Having come under fire for selling a product that may be causing diabetes, Robertson said, "they wanted to position themselves so that they could say they were actually helping do research on diabetes."

The bulk of the industry's effort would go to continuing the public-relations battle. By concentrating its efforts on the FDA report, Tatem would describe in memos and presentations, the Sugar Association would actually lose the next battle in the war. The industry had been confident that George McGovern's committee, which had held the 1973 hearings on sugar, "would self destruct" in 1977, and so the Sugar Association had focused its attention on the FDA. But the committee survived long enough to publish a report, *Dietary Goals for the United States,* in January of that year. McGovern would describe the report in a press conference as "the first comprehensive statement by any branch of the Federal Government on risk factors in the American diet." The committee's report would focus primarily on getting Americans to eat less fat, but it would also recommend that the nation reduce its sugar consumption by 40 percent, a number in tune with George Campbell's estimate of the threshold at which populations begin to manifest diabetes epidemics. The sugar industry was taken by surprise.

Tatem told Sugar Association members that they had "hammered away" at McGovern's committee afterward, using the FDA report "as our scientific bible," but McGovern ("or more likely his staff," according to Tatem) wasn't impressed and wouldn't budge off the 40 percent number. It stayed in a revised edition of the *Dietary Goals,* which was published at the end of 1977. "The weight given to the consideration of sugar's relationship to obesity and disease is a matter of judgment," McGovern wrote to Tatem in a letter, "and I believe we have been prudent in our judgment."

After the McGovern report, though, the Sugar Association and the industry carried the day. In 1980, the U.S. Department of Agriculture released the first edition of its "Dietary Guidelines," drafted by a small committee led by Mark Hegsted, who had spent his entire career working in Fred Stare's department at Harvard. Hegsted later said that he had relied on Ed Bierman's 1979 review in the *American Journal of Clinical Nutrition* to decide how to phrase the sugar recommendations, and Bierman had been confident that sugar was harmless.

"Contrary to widespread opinion," the "Dietary Guidelines" said, "too much sugar does not seem to cause diabetes." It then advised that we "avoid too much sugar," without bothering to define what was meant by "too much." In the second edition of the guidelines, published in 1985, the USDA (with Fred Stare now a member of the guidelines advisory committee) was still advising Americans to avoid too much sugar, but had now dropped the caveat on the diabetes-sugar connection. Instead, it stated unambiguously that "too much sugar in your diet does not cause diabetes," even though much of the significant research published in the intervening years had come out of the USDA's own Carbohydrate Nutrition Laboratory and supported the notion that sugar consumption was, indeed, a cause of diabetes, and that even "modest" amounts of sugar could increase the risk of heart disease in a significant proportion of the population.

In 1986, the FDA returned to the question of whether sugar should be generally recognized as safe. Three FDA administrators, led by Walter Glinsmann (who would later become a consultant for the Corn Refiners Association), now took up the job that the SCOGS committee had left off in 1976. After reviewing the evidence once again, these FDA administrators determined that "no conclusive evidence demonstrates a hazard to the general public when sugars are consumed at the levels that are now current."

The FDA assessment then became the official government position on sugar, its logic and conclusions echoed in a series of official reports on diet and health that came after — particularly the 1988 *Surgeon General's Report on Nutrition and Health* and the 1989 National Academy of Sciences report *Diet and Health,* which are the two seminal documents on the subject in the last half-century, and even reviews by the Institute of Medicine as late as 2005. All of these official documents focused on fat as the root of dietary evils: The "disproportionate consumption of food high in fats," according to the Surgeon General's report, played a prominent role in five of the ten most common causes of death and thus could be held chiefly responsible for two-thirds of the 2.1 million deaths in the United States that year. All repeated the FDA's conclusion that the evidence linking sugar to chronic disease was inconclusive, and then effectively equated "inconclusive," as the Sugar Association did, with "nonexistent." (As of March 2016, the Sugar Association Web site was still misquoting the FDA report to make that point.)

All of these seminal reports also ignored a second caveat that accompanied the 1986 FDA review of sugar: the FDA report had concluded that sugar was likely to be harmless "when sugars are consumed at the levels that are now current." As Walter Glinsmann would later explain, any substance could be harmful if taken at too high a dose, so the levels at which a substance is taken in a drug or consumed in a diet are key. (This logic was contrary to that used by the SCOGS panels, for instance, in condemning cyclamates and saccharin — the dosage necessary to induce cancer in an animal model was considered irrelevant — but the FDA and Glinsmann's committee invoked it with sugar nonetheless.)

In their 1986 report, Glinsmann and his colleagues estimated the levels at which sugar was currently consumed to be forty-two pounds of sugar per person per year, or the equivalent every day of the amount of sugar in eighteen ounces — a can and a half — of Coke or Pepsi. This was only slightly more than half of what the USDA was estimating at the time — seventy-five pounds per capita — and significantly less than half (44 percent) of what the USDA estimated we were consuming by the early twenty-first century, ninety pounds per capita. Even the most ardent critics of sugar would probably be content if Americans consumed only forty-two pounds of added sugar and high-fructose corn syrup each year on average, but the evidence suggests we consume significantly more.

In 1989, the British Committee on Medical Aspects of Food Policy (commonly known as COMA) released the British government's first official assessment of the health aspects of sugar, a report entitled *Dietary Sugars and Human Disease.* The committee that authored the report was composed of a dozen of the leading nutritionists, biochemists, and physiologists in the U.K., led by a diabetes specialist named Harry Keen, who had received funding from the sugar industry throughout the 1970s.

The British report clearly manifested the conflict between the urge to exonerate sugar — based on, if nothing else, what the FDA and hence the surgeon general's office and the National Academies of Science were now claiming — and the scientific evidence itself. Keen and his colleagues acknowledged that chronic consumption of sugar at the levels the British public seemed to be consuming at the time (roughly equivalent to the seventy-five pounds per capita the USDA was then estimating for American consumption) could induce, as Yudkin had proposed, a cluster of metabolic abnormalities associated with elevated levels of triglycerides and thus heart disease, diabetes, hypertension, and obesity. It acknowledged that some significant portion of the population was sensitive to sugar and other carbohydrates. But it then concluded that sugar "played no causal role" in these diseases. The one major caveat in the British report was that individuals with elevated levels of triglycerides — a proportion that today, for instance, might constitute as much as half of the adult population in the United Kingdom or the United States — would be best served by restricting their consumption of sucrose and other "added sugars" to twenty to forty pounds per year, or roughly what the British were consuming per capita in the early years of the Victorian era — almost two hundred years earlier.

CHAPTER 9
WHAT THEY DIDN'T KNOW

> I wish there were some formal courses in medical school on Medical Ignorance; textbooks as well, although they would have to be very heavy volumes.
>
> LEWIS THOMAS, "Medicine as a Very Old Profession," 1985

Over the past four hundred years, thinking on the scientific method has distilled the concept down to two words: "hypothesis" and "test." If we want to establish reliable knowledge — that what we think is true really is — this is the process that must be followed. In the words of the philosopher of science Karl Popper, "The method of science is the method of bold conjectures and ingenious and severe attempts to refute them." The bold conjectures are the hypotheses, and they are the relatively easy part of science. The ingenious and severe attempts to refute them are the experimental tests —

the hard part. This is what takes time, effort, and money, and often prohibitive amounts of each.

Nutrition hypotheses are particularly challenging because they're often about how foods or constituents of foods or dietary patterns influence our pursuit of a long and healthy life. The hypothesis addressed in this book, for instance, is that sugar is the dietary trigger of obesity and diabetes and, if so, the diseases such as heart disease that associate with them. But this hypothesis is ultimately about what happens to us over decades — the time it takes chronic diseases to manifest themselves — and not months, as is the case, say, with vitamin-deficiency diseases like scurvy or beriberi.

In the late 1960s, when administrators at the National Institutes of Health considered doing a trial that would test the hypothesis that dietary fat causes heart disease and thus, ultimately, the shortening of our lives, they concluded that it would require perhaps a hundred thousand subjects and would cost at least one billion dollars. And they were justifiably concerned that such a study still couldn't be trusted to give a reliable and definitive result. (That's why replication, ideally by independent investigators, is also considered key to the scientific process: a necessary step before a hypothesis is accepted as likely to be true.) So such a study was never undertaken.

What happened after that tells us a lot about the particular pitfalls of nutrition science and public-health policy and how they interact. Instead of the billion-dollar test of the dietary-fat hypothesis, the NIH invested a quarter-billion dollars in two trials that tested variations on the same theme, or links in a hypothetical chain of reasoning. The first trial would test the supposition that men with high cholesterol levels who were told to eat a low-fat diet (and also took blood-pressure medication and received counseling to quit smoking, if either of these was necessary) would live longer than men who weren't. The results of this study were published in 1982 and failed to confirm the hypothesis. The men on the low-fat diet suffered more deaths than the men who were left to their own devices. (The investigators refused to believe that a low-fat diet could be harmful, and certainly not the smoking cessation, so they concluded, questionably, that the blood-pressure medication had unforeseen side effects and caused more deaths than it prevented.) The second trial tested the hypothesis that a cholesterol-lowering medication given to men with very high levels of cholesterol would lengthen their lives, compared with men who took no such medication. The results of this study, published in 1984, indicated that the medication helped, albeit just barely.

The authorities at the National Institutes of Health then took what amounts to a leap of faith. ("It's an imperfect world," as one of the NIH administrators later phrased it. "The data that would be definitive are ungettable, so you do your best with what is available.") Concerned, as they were, that hundreds of thousands of Americans were dying of heart disease yearly, they assumed that if a drug that lowered cholesterol would extend the lives of men with very high cholesterol, then a diet that also lowered cholesterol would do the same for all the rest of us. Equally important, they assumed that the benefit of communicating this leap of faith on a nationwide scale was worth the risks. In 1984, attended by considerable controversy, they initiated a massive public-relations campaign to induce every American over the age of two to eat a low-fat diet. We've been living with the consequences ever since.

Had scientific progress stopped there, we wouldn't know whether the leap of faith was justified. But we do. The NIH eventually spent between half a billion and a billion dollars, depending on the estimate, testing the hypothesis that a low-fat diet would prevent chronic disease in women and bestow on them a longer life. The authorities involved had little doubt that it would, and were responding to political pressure to include women in medical trials; women had been underrepresented until then. The trial, known as the Women's Health Initiative, was launched in the early 1990s, and the results were reported in 2006. Once again, it failed to confirm the hypothesis. The roughly twenty thousand women in the trial who had been counseled to consume low-fat diets (and to eat more fruits, vegetables, and whole grains, and less red meat) saw no health benefits compared with the women who had been given no dietary instructions whatsoever.

Once again, the researchers involved and the public-health authorities chose *not* to perceive this negative result as reason to question their belief that fat causes heart disease and that low-fat diets will prevent it. Rather, they chose to assume that the trial — the largest such randomized trial ever done — simply failed to get the right answer, or would have gotten the answer they expected ("statistically significant," in the scientific jargon) had the study lasted longer or included more subjects, or had the women in the trial done a better job of adhering to a low-fat diet. These authorities had now spent decades (nearly half a century, in the case of the American Heart Association) telling us that dietary fat was killing us. Thus they found it easier to accept, or at least easier to communicate, the notion that the study had failed (or almost but not quite succeeded) than that their preconceptions about diet and the dietary advice they had been giving, based largely on that initial leap of faith, had been incorrect.

Often in science, repeated tests of a hypothesis result not in its disproof but in less and less reason to believe it's true. That was the case with the dietary-fat theory. In 1987, as we've seen, in the midst of the government's public-health campaign — i.e., the leap of faith — a supposedly definitive *Surgeon General's Report on Nutrition and Health* had claimed that two in every three of the two million yearly deaths in the United States could be blamed chiefly on "the disproportionate consumption of food high in fats," and that "the depth of the science base . . . is even more impressive than that for tobacco and health in 1964." A quarter century later, the most authoritative review of the evidence — from an international organization known as the Cochrane Collaboration — claimed that no health benefits derived from eating a diet low in fat, although the evidence "suggest[ed]" a small benefit if a diet high in fat replaced saturated fat with polyunsaturated fat. The leap of faith had turned out to be, well, a leap of faith.

At the core of all nutrition controversies is a simple fact: the requirements of public-health policy and the requirements of good science can be mutually exclusive. When large numbers of Americans are dying from diet-related diseases, leaps of faith can be justified if the odds seem good that they will save lives. In fact, it may seem irresponsible not to take such steps. But leaps of faith are incompatible with the institutionalized skepticism required to do good science, and the process of rigorously and repeatedly testing our beliefs to establish whether or not they're true. Public-health authorities will talk about not having the time to gather "definitive scientific evidence," because they believe they have to act. Scientists will argue that the absence of definitive scientific evidence means that we don't know what the truth is and, therefore, how to act. And they may both be right. In

1999, when I first started my investigations into these nutrition controversies for the journal *Science,* the then director of the NIH's office of disease prevention, William Harlan, put it this way: "We're all being pushed by people who say, 'Give me the answer. Is it or isn't it?' They don't want the answer after we finish a study in five years. They want it now. No equivocation . . . [and so] we constantly get pushed into positions we may not want to be in and cannot justify scientifically."

One danger here, of course, is that once we insist or pretend that we know the answer based on premature or incomplete evidence (even if we're pushed against our will to take such stands), we're likely to continue to insist we're right, even when evidence accumulates to the contrary. This is a risk in any human endeavor. When Francis Bacon pioneered the scientific method almost four hundred years ago, he was hoping to create a methodology of critical or rational thinking that would minimize this all-too-human characteristic of avoiding evidence that disagrees with any preconceptions we might have formed.* Without

* "The human understanding," wrote Bacon, "once it has adopted opinions, either because they

rigorous tests, as many as necessary, beliefs and preconceptions will persevere because it's always easier to believe that a single test has been flawed, or even a few of them, than it is to accept that our belief had been incorrect. The scientific method protects against this tendency; it does not eradicate it.

In 1969, John Yudkin discussed this conflict in the context of nutrition research and, specifically, the challenges of establishing reliable knowledge about sugar and chronic disease. Speaking at a symposium in London, Yudkin acknowledged that none of the existing research on sugar could be considered definitive. No one had yet tested the actual hypotheses that were being debated. Scientists had tested the hypothesis that sugar consumption caused chronic disease in rats, because they could do those experi-

---

were already accepted and believed, or because it likes them, draws everything else to support and agree with them. And though it may meet a greater number and weight of contrary instances, it will, with great and harmful prejudice, ignore or condemn or exclude them by introducing some distinction, in order that the authority of those earlier assumptions may remain intact and unharmed."

ments: they could feed the rodents sugar-rich diets, or not, and see what happened over the lifetime of a rat. But it wasn't a human's lifetime. They had no idea whether rats were good models for humans. Moreover, as other researchers had implied at the same conference, they couldn't even know if the rats they used were good models for other rats, since some of the observations were what researchers would call "strain specific." Eating sugar seemed to shorten the lives of some strains of rats but not others.

The kind of randomized controlled trials over the course of ten or twenty years that would truly test the hypothesis that sugar caused heart disease or diabetes, as Yudkin noted, were no different from the kind the NIH was then considering and would soon reject for the dietary-fat/cholesterol hypothesis. Such trials were certainly far beyond the budget of any single researcher or even collaboration of researchers; they required that the National Institutes of Health or the Medical Research Council in the U.K. or some other government agency create a concerted program to test the idea. Without that, researchers would do what they could afford: study rats or primates, or study a few dozen human subjects for weeks or a few months, and see what happened. "It would be just as great a mistake to dismiss the results of such experiments as valueless because of these limitations," Yudkin said, "as to accept them uncritically as answering questions relating to long-term diets in all persons."

In 1986, with the perceived FDA exoneration of sugar, the public-health authorities and the clinicians and researchers studying obesity and diabetes had come to a consensus that type 2 diabetes was caused by obesity, not sugar, and that obesity itself was caused merely by eating too many calories or exercising away too few. By this logic, the only means by which a macronutrient could influence body weight was its caloric content, and so, calorie for calorie, sugar was no more fattening than any other food, and thus no more likely to promote or exacerbate diabetes. This was what the sugar industry had been arguing and embracing since the 1930s. It was what Fred Stare of Harvard had in mind when he said publicly that he would prefer to get his calories from a martini than from a dessert.

A more nuanced perspective, one nourished by scientific progress, would be that if two foods or macronutrients are metabolized differently — if glucose and fructose, for instance, are metabolized in entirely different organs, as they mostly are — then they are likely to have vastly different effects on the hormones and enzymes that control or regulate the storage of fat in fat cells. One hundred calories of glucose will very likely have an entirely different effect on the human body from one hundred calories of fructose, or fifty calories of each consumed together as sucrose, despite having the same caloric content. It would take a leap of faith to assume otherwise.

Nutritionists had come to assume that a hundred calories of fat had a different effect from a hundred calories of carbohydrate on the accumulation of plaque in coronary arteries; even that a hundred calories of saturated fat would have an entirely different effect from a hundred calories of unsaturated fat. So why not expect that macronutrients would have a different effect on the accumulation of fat in fat tissue, or on the phenomena, whatever they might be, that eventually resulted in diabetes? (Insulin resistance and hyper-insulinemia, as Rosalyn Yalow and Solomon Berson, among others, had suggested in the 1960s, seemed to be a very likely bet.) But obesity and diabetes researchers, as we've seen, had come to embrace the mantra that "a calorie is a calorie"; they would repeat it publicly when they were presented with the idea that there was something unique about how the human body metabolizes sugar that sets it apart from other carbohydrates. The long-held view was based on the state of the science in the early years of the twentieth century, and to cling to it required a willful rejection of the decades' worth of relevant revelations in the medical sciences that had come since.

By the 1980s, biochemists, physiologists, and nutritionists who specialized in the study of sugar or in the fructose component of sugar had come to consistent conclusions about the short-term effects of sugar consumption in human subjects, as well as the details of how sugar is metabolized and how this influences the body as a whole. The glucose we consume — in starch or flour, or as half of a sugar molecule — will be used directly for fuel by muscle cells, the brain, and other tissues, and can be stored in muscles or the liver (as a compound called glycogen), but the fructose component of sugar has a much different fate. Most of it never makes it into the circulation; it is metabolized in the liver. The metabolic pathways through which glucose passes when it is being used for fuel — in both liver and muscle cells — involve a feedback mechanism to redirect it toward storage as glycogen when necessary. This is the case with fructose, too. But the metabolism of fructose in the liver is "unfettered by the cellular controls," as biochemists later put it, that work to prevent its conversion to fat. One result is the increased production of triglycerides, and thus the abnormally elevated triglyceride levels that were observed in many research subjects, though not all, when they ate sugar-rich diets.

While cardiologists and epidemiologists were debating whether elevated triglycerides actually increased the risk of heart disease (in the process, challenging their own beliefs that cholesterol was key), biochemists had come to accept that sucrose was "the most lipogenic" of carbohydrates — as even Walter Glinsmann, author of the FDA report on sugar, would later acknowledge — and that the liver was the site of this fat synthesis.* The Israeli biochemist

* In 1916, when Harold Higgins of the Carnegie Institute published the first studies on how rapidly we metabolize different carbohydrates, he had made this same observation. Fructose (and sometimes galactose) "shows a tendency or preference to change to fat in the body, while glucose

Eleazar Shafrir would describe this in the technical terminology as "the remarkable hepatic lipogenic capacity induced by fructose-rich diets." It was also clear from the short-term trials in humans that this happened to a greater extent in some individuals than others, just as it did in some species of animals and not others. In human studies, subjects who had the highest triglycerides when the trials began tended to have the greatest response to reducing sugar intake, suggesting (but not proving) that the sugar was the reason they had such high triglycerides in the first place. These same individuals also tended to see the greatest drop in cholesterol levels when they were put on low-sugar diets.

There were other interesting vagaries in how both humans and animals in these experiments responded to sugar that these researchers would have liked to explore further, but government funding for this kind of research was increasingly hard to come by in the latter half of the 1980s. Young women, for instance, seemed relatively resistant to this triglyceride-raising effect of sugar, whereas older and particularly

---

tends to change to glycogen [the storage form of carbohydrate] and be stored as such."

post-menopausal women responded just like men. The researchers doing these studies wondered if this could explain why younger women seemed relatively immune to heart disease, but all they could do is speculate.

Subjects who responded with elevated triglycerides to sugar-rich diets also tended to manifest a phenomenon known as glucose intolerance when they consumed carbohydrates: their blood-sugar level over the next few hours would rise higher than it should have. This suggested that the cells of these individuals might also be relatively resistant to the action of insulin in working to keep blood sugar under control. But it wasn't clear why this happened, particularly since the sugar itself was being metabolized in the liver and the fructose component of sugar was not even stimulating the pancreas to secrete insulin. In the early 1970s, Aharon Cohen and his Israeli colleagues had reported that these individual responses were very likely determined by genetic proclivities and that they were linked to the eventual onset of diabetes, at least in rats. Cohen and his colleagues had bred together lean rats that were otherwise healthy, except for this phenomenon of becoming glucose-intolerant on sugar-rich diets. Then they had taken the offspring of these rats, the ones that were also glucose-intolerant when they ate sugar, and bred them together. Within three generations, the progeny would become diabetic upon eating sugar, not just glucose-intolerant. Whether this meant the same thing happened in humans, and whether it explained why some of us get diabetic while eating no more sugar than others who don't, was something neither Cohen nor anyone else could answer.

In 1986, when Walter Glinsmann and his colleagues compiled the final FDA report on sugar, they discussed many of these findings, and then chose to take the absence of definitive evidence on *long-term* effects of sugar consumption as sufficient reason to conclude that sugar was generally recognized as safe. By then, the great majority of researchers and clinicians thinking about heart disease had come to accept that fat was the problem, not sugar, and so they did, indeed, generally consider sugar to be safe. That didn't mean it *was* safe, only that this was what most authorities who were expected to have an informed opinion in the 1980s believed.

Researchers who argued otherwise, such as Yudkin, Walter Mertz, and Sheldon Reiser at the USDA Carbohydrate Nutrition Laboratory, were assumed to be biased or bad scientists or, like Yudkin, overly invested in a quack hypothesis. The kinds of tests necessary to answer the question definitively had never been done, and Glinsmann and his co-authors had offered up no suggestions about whether they should be. In fact, their charge in compiling the FDA report did not include specifying where more research was necessary, and so they didn't.* Dietary fat had been proclaimed the dietary cause of heart disease, and the government and health organizations would now commit themselves to getting Americans to eat low-fat diets.

The context would soon change on the science of sugar, but not before two other developments that influenced how the nutritional authorities perceived it and, perhaps more important, how the public perceived *and* consumed it. Throughout the twentieth century, diabetes specialists and nutritionists had assumed that if any component of the food we ate caused or exacer-

* Twenty-five years later, when I asked Walter Glinsmann, who was then consulting for the Corn Refiners Association, what research could be done to resolve the sugar question definitively, he refused to answer the question.

bated diabetes, either it had to make us fatter (dietary fat by the 1980s was widely touted as the prime suspect for that, because of its particularly dense calories) or it had to put a unique strain on the insulin-secreting cells of the pancreas. Even the British researcher Peter Cleave had assumed this to be true, and it had strongly influenced his thinking in the 1960s, when he was arguing that refined grains and sugars were the causes of obesity and diabetes and their associated chronic diseases.

If this was true, then the key factor in how sugar or any carbohydrate influenced diabetes status would likely be how quickly these foods were digested into their component carbohydrates, such that the glucose could be released into the circulation and result in a rise in blood sugar. This concept came to be known as the "glycemic index." It was pioneered in the late 1970s by researchers at Oxford University, and it supported the notion that Cleave had been right, at least in this one sense. The more refined or processed a carbohydrate, and the less fat and fiber accompanying it to slow its digestion, the greater the blood-sugar response, and thus the more insulin required to metabolize it; or, as Cleave might have phrased it, the greater the strain on the pancreas. For the glycemic index, the Oxford researchers established a reference value of 100 when subjects drank a solution of glucose and water alone. Corn flakes rated 80, white rice 72, white bread 69, apples 39, and ice cream (with its high fat content) only 36.

The initial publications on the glycemic index sparked a surprisingly acrimonious controversy about its ultimate value. One obvious problem was that the blood-sugar response to consuming any specific food would differ significantly from person to person and be strongly influenced by the meals in which that food was consumed — how much fat, protein, and fiber were contained in the other foods in the meal. Another problem was that a food rich in fat, and even saturated fat — ice cream being the prime example — would have a low glycemic index because of the fat content and so appear, by this measure, to be healthy. Many nutritionists and researchers concerned about obesity, diabetes, and heart disease and convinced that dietary fat was the culprit found this to be an unacceptable conclusion. Still, the concept of the glycemic index would slowly come to be embraced by the diabetes community as a useful measure of what foods diabetics could or could not eat, or how they had to modulate their insulin doses if they did.

An unintended consequence of the glycemic index is that it made sugar seem healthy, even for diabetics. Because most of the fructose we consume never makes it through the liver to show up in the circulation as blood sugar, fructose barely registers in the glycemic index. As a result, sugar (now sucrose *and* high-fructose corn syrup, as we'll discuss shortly) has a relatively low glycemic index — only half of it, the glucose, raises blood sugar. This made fructose appear to be an ideal sweetener for diabetics, and sugar itself of little concern. There was no reason, therefore, "for diabetics to be denied foods containing sucrose," as University of Minnesota researchers concluded in a 1983 article in *The New England Journal of Medicine.* By 1986, this was the official position of the American Diabetes Association as well.

This helps to explain the rise in total caloric-sweetener consumption — in the consumption of sugars that contain fructose, specifically sucrose and high-fructose corn syrup (HFCS) — that began in the 1980s and paralleled the latest incarnations of the obesity and diabetes epidemics. We went from the first half of the 1970s, during which sugar was vilified and per capita sugar consumption actually dipped, to the 1980s, which saw the beginning of the first significant increase in total intake since the Great Depression. In 1999, when 150 pounds of sugar and HFCS were being sold in the United States for every man, woman, and child in the country, this was a third more than had been available a quarter century earlier (113 pounds). Depending on how it's calculated (what proportion of the sugar and HFCS sold is then actually consumed), by 1999 we were now eating and/or drinking from two to three times the dose of sucrose and HFCS that Glinsmann and his FDA colleagues had officially defined as safe just thirteen years earlier.

The upturn began after the sugar industry's successful public-relations campaign and shortly before the exoneration of sugar by the FDA. It coincided with the introduction of high-fructose corn syrup into the food supply, and particularly what is known as HFCS-55 — the aforementioned mixture of 55 percent fructose and 45 percent glucose that had been created to be indistinguishable from sucrose when used to sweeten Coca-Cola or Pepsi.* By 1984, it

* The fructose and glucose in HFCS are not

had replaced sucrose in both these soft drinks, largely because it was cheaper and, thanks to government legislation passed by the Reagan administration, could be trusted to remain cheaper. It also came in the form of a syrup that was particularly convenient for the beverage industry. From 1984 through the end of the century, caloric-sweetener consumption steadily rose as HFCS first replaced a fair share of the sucrose we were consuming, and then kept climbing.

Multiple possible explanations exist for why this happened, including the fact that the public-health authorities were now telling Americans that fat was what made them fat and implying that sugar was effectively harmless, as long as we didn't overdo it. (By the mid-1990s, even the American

---

bound together as they are in sucrose, which has led some researchers to suggest that HFCS may be inherently more harmful. This may be less relevant than these researchers believe, though, because much of the sucrose in the food supply, and particularly in soft drinks — estimated in the 1970s at perhaps 50 percent — ends up as "invert sugar," in which the fructose and glucose have also been broken apart (hydrolyzed) by the time we consume it.

Heart Association was recommending we have sugar candies for snacks, rather than foods that contained saturated fat.) Another simple explanation is that the corn refiners went out of their way to promote HFCS as something other than sugar. They referred to their product as "fructose," as though that's all it was, and then they referred to "fructose" as "fruit sugar," making it seem inherently healthy. With the American Diabetes Association and diabetes specialists now suggesting that fructose is an ideal sweetener on the basis that it doesn't raise blood sugar or require insulin to be metabolized, this made HFCS seem ideal as well.

It's difficult to imagine that we simply failed to realize that the HFCS we were now consuming in our soft drinks and juices and an ever-increasing number of processed foods and baked goods was, indeed, just another form of glucose and fructose and thus, in effect, sugar, but that is what happened. The corn refiners had succeeded in muddying the difference.* HFCS became

* When I began the research and reporting for my first book on nutrition in the early 2000s, even many of the researchers I interviewed either believed that HFCS was fructose alone or didn't know that sucrose was half fructose. Because these

the sweetener of choice in a host of products that were now portrayed as uniquely healthy — sports drinks like Gatorade; bottled teas infused with ginkgo biloba, ginseng, or other exotic herbs; low-fat yogurts — and exploded in popularity at the time. The manufacturers could acknowledge in the list of ingredients that the primary source of calories came from high-fructose corn syrup without alerting consumers that this was just another form of sugar, and that they might get even fatter and perhaps more likely become diabetic because of it. As it turned out, we did get fatter *and* more diabetic. The question, of course, is whether that is a coincidence or an instance of cause and effect.

In the late 1980s, the context of the science itself began to shift radically. The biochemistry of how the liver metabolizes fructose had been well worked out, and why sugar consumption would be expected to elevate triglycerides in the bloodstream. That was

---

researchers tended to be either epidemiologists who study populations, or physicians who backed chronic diseases, they didn't have the nutrition or biochemistry background necessary at the time to be aware of these simple facts.

not controversial. But the medical context in which it would be understood — or, more precisely, *should* be understood — would change. A series of developments in our understanding of heart disease and diabetes began to take the spotlight away from the cholesterol/dietary-fat connection and shine it on the carbohydrate content of the diet.

The medical research community came to recognize that insulin resistance and a condition now known as "metabolic syndrome" is a major, if not *the* major, risk factor for heart disease and diabetes. Before we get either heart disease or diabetes, we first manifest metabolic syndrome. The CDC now estimates that some seventy-five million adult Americans have metabolic syndrome.

The very first symptom or diagnostic criterion that doctors are told to look for in diagnosing metabolic syndrome is an expanding waistline. This means that if you're overweight or obese — as two-thirds of American adults are — there's a good chance that you have metabolic syndrome; it also means that your blood pressure is likely to be elevated, and you're glucose-intolerant and thus on the way to becoming diabetic. This is why you're more likely to have a heart attack than a lean individual — although lean individuals can also have metabolic syndrome, and those who do are more likely to have heart disease and diabetes than lean individuals without it.

Metabolic syndrome ties together a host of disorders that the medical community typically thought of as unrelated, or at least having separate and distinct causes — getting fatter (obesity), high blood pressure (hypertension), high triglycerides, low HDL cholesterol (dyslipidemia), heart disease (atherosclerosis), high blood sugar (diabetes), and inflammation (pick your disease) — as products of insulin resistance and high circulating insulin levels (hyperinsulinemia). It's a kind of homeostatic disruption in which regulatory systems throughout the body are misbehaving with slow, chronic, pathological consequences everywhere.

The research on metabolic syndrome dates back to the early 1950s and ties together Rosalyn Yalow and Solomon Berson's revelation that both the obese and type 2 diabetics are insulin-resistant with the science Yudkin invoked in 1963 to argue that sugar consumption was the most likely dietary cause of heart disease. Virtually all of these disorders could be generated by feeding sugar to laboratory animals, as Yudkin pointed out, and many by feeding sugar to humans. The Stanford University endocrinologist Gerald Reaven and his collaborators deserve the credit for much of the additional science, and for then getting the medical community to pay attention, a considerable feat. Reaven's argument would be a variation on Yudkin's: that heart disease and diabetes are associated with a common set of metabolic and hormonal disruptions, including obesity, and that elevated cholesterol levels may be the least of them. Reaven implicated all carbohydrates in the disease state. Unlike Yudkin, he wasn't considered a zealot who argued that sugar was toxic and saturated fat was not.

In 1987, Reaven discussed the emerging science of metabolic syndrome at a conference on diabetes prevention hosted by the National Institutes of Health. The researchers and clinicians in attendance acknowledged that the science was compelling, but they also wished, as one NIH administrator said at the time, that "it would go away, because nobody knows how to deal with it." They had come to believe that fat was bad for the heart and that too much protein could put an unhealthy strain on the kidneys. Now Reaven was bringing back the notion that carbohydrates were bad. "We have to eat something," the NIH official said, but what would be left?

The following year, Reaven gave the prestigious Banting Lecture at the annual meeting of the American Diabetes Association. He described the evidence supporting what he had come to call "Syndrome X" (metabolic syndrome). As Reaven described it, the condition of being resistant to insulin — the key defect in metabolic syndrome — is the underlying cause of type 2 diabetes. Not everyone with insulin resistance becomes diabetic, however; some continue to secrete sufficient insulin to overcome their bodies' resistance to the hormone. And this hyperinsulinemia in turn has deleterious effects throughout the human body, including causing heart disease by raising triglyceride levels and blood pressure, lowering levels of HDL cholesterol, and further exacerbating the insulin resistance. It's a vicious cycle in which secreting too much insulin can cause insulin resistance, and insulin resistance will cause the body to secrete still more insulin. Diabetes and heart disease are likely to follow. Getting ever fatter may be a cause, but it could be a result as well.

Over the years, as the research on metabolic syndrome has accumulated, it has generated an ever-growing list of metabolic and hormonal abnormalities that accompany insulin resistance and are thus found in the obese, and which precede both heart disease and diabetes. These include large numbers of LDL particles in the circulation (not the cholesterol itself, but the particles that carry the cholesterol) and elevated blood levels of uric acid, a precursor of gout. They also include a state of chronic inflammation, marked by a high concentration in the blood of a protein known as C-reactive protein and other inflammatory molecules.

Metabolic syndrome changes the vocabulary that physicians use when they discuss a patient's risk of heart disease. High cholesterol isn't among the cluster of metabolic abnormalities, nor is elevated LDL cholesterol, the "bad" cholesterol. Rather, the key factors are high triglycerides, low HDL cholesterol, high blood pressure, overweight, glucose intolerance, and, more than anything, the condition of being insulin-resistant and thus oversecreting insulin, day in and day out. All of these abnormalities happen to be related to the carbohydrate content of the diet, not to the fat content.

The ultimate question, though, is what causes the insulin resistance? What sets off this vicious cycle? Since the early 1960s, many researchers and clinicians have been willing to assume that it's obesity, or at least excess fat accumulation, for the same reason they assumed obesity caused diabetes — the two are so closely associated. But this doesn't explain how lean people can also be insulin-resistant (or diabetic), so sedentary behavior is often invoked to explain metabolic syndrome in these cases. Both are a way to reconcile the presence of insulin resistance in obesity while still blaming obesity itself on more calories consumed than expended. These assumptions were never rigorously tested, but they seemed reasonable and so they were embraced.

One of the interesting side effects of the research on the glycemic index, though, and then the slow acceptance of insulin resistance and hyperinsulinemia as both precursors and drivers of heart disease and diabetes, is that the number of researchers studying sugar and its fructose component began to increase again in the late 1980s. This wasn't because the researchers were particularly concerned that sugar was bad for us. Rather, some started studying fructose because it was seen as a potentially ideal sweetener for diabetics, as the American Diabetes Association was saying, and some because fructose presented a means of comparison with glucose for laboratory studies of metabolism — one had an immediate effect on blood sugar and insulin secretion (glucose), and the other did not (fructose).

Some researchers began studying fructose because researchers in Reaven's laboratory at Stanford demonstrated that the easiest way to cause the symptoms of insulin resistance and thus metabolic syndrome in laboratory rats and mice was to feed them large amounts of fructose. As Reaven would later explain, they started feeding diets that were mostly fructose to their rats because they were curious about the recommendations from the American Diabetic Association. The Stanford researchers very quickly found that they had "a marvelous model" for the metabolic syndrome they were studying in humans — high triglycerides, high insulin levels (hyperinsulinemia), insulin resistance, even high levels of uric acid.

Some researchers began studying sugar because they were interested in why fat accumulates in the liver. The first reports linking fatty liver disease to obesity in humans date to 1950 and a Kansas physician named Samuel Zelman, who suggested that the carbohydrate load consumed by his obese patients might somehow be responsible. (He was motivated to study the subject, he wrote, by a patient who happened to be an aide in his hospital and "ingested the contents of 20 or more bottles of coca-cola per day.") The first case reports in the literature diagnosing fatty liver disease in adults who had no history of alcohol consumption — hence, *nonalcoholic* fatty liver disease, or NAFLD — date to 1980 and, in children, to 1984. The condition is indistinguishable from the fatty liver disease that alcohol is known to cause. Its presence in adults who don't drink and in children was explained by the fact that these patients were almost invariably obese and had high triglycerides. In other words, they had metabolic syndrome.

Today one in every ten adolescents is thought to have nonalcoholic fatty liver disease, as are an estimated seventy-five million adults (perhaps not coincidentally, the same number as are estimated to have metabolic syndrome). The condition has now been diagnosed in infants. It's clearly another epidemic. Some clinicians dealing with NAFLD assume it's caused by obesity; others have wondered what aspect of modern diets or lifestyles could uniquely work to make fat accumulate in the liver. Because

NAFLD also very closely associates with metabolic syndrome and insulin resistance, one possibility is that it's the accumulation of fat in the liver that actually causes the insulin resistance that is at the heart of metabolic syndrome. This is what many researchers who study insulin resistance believe today, and what the latest evidence suggests. But why does fat accumulate in the liver? Some of the researchers trying to answer that question are studying sugar, because fructose is metabolized in the liver and is highly lipogenic (fat-producing).

Since the 1990s, these researchers have established certain findings unambiguously. First, feed animals enough pure fructose or enough sugar (glucose and fructose) and their livers convert much of the fructose into fat — the saturated fat palmitic acid, to be precise, which is the one that *supposedly* gives us heart disease when we eat it, by raising LDL cholesterol. The biochemical pathways involved are clear and not particularly controversial. Feed animals enough fructose for long enough and this fat accumulates in the liver, causing the kind of fatty liver seen in obese children and adults. The fat accumulation accompanies insulin resistance, first in the liver and then in other cells as well, resulting in metabolic syndrome, at least in laboratory animals.

These researchers say the metabolic effects of consuming sugar or fructose can happen in as little as a week if the animals are fed huge amounts of it — almost 70 percent of the calories in their diets. The effects may take several months to appear if the animals are fed something closer to what humans in America actually consume — around 20 percent of the calories in their diet. Stop feeding them the sugar, in either case, and the fatty liver goes away, and with it the insulin resistance. In a 2011 study in which twenty-nine rhesus monkeys were given the opportunity to drink a fructose-sweetened beverage along with their usual monkey chow, every last one of them developed "insulin resistance and many features of the metabolic syndrome" within a year, and four had progressed to type 2 diabetes.

Researchers have obtained similar results with humans (albeit without going so far as to give them diabetes), but they have typically done the experiments only with fructose. Luc Tappy at the University of Lausanne in Switzerland began studying fructose in the mid-1980s because he was "fascinated by the very peculiar metabolism of fructose, [that] it's readily metabolized without the need of insulin." When Tappy fed his human subjects the equivalent of the fructose in eight to ten cans of Coke or Pepsi a day — a "pretty high dose," as he says — their livers would start to become insulin-resistant and their triglycerides would elevate in just a few days. With lower doses, the same effects would appear but only if the experiment ran for a month or more.

Despite the steady accumulation of research implicating sugar and fructose in the accumulation of fat in the liver and insulin resistance, every experiment can still be easily criticized as falling short of being conclusive — just as Walter Glinsmann and his FDA co-authors suggested in 1986. The studies with rodents aren't necessarily applicable to humans. And the kinds of studies that Tappy did — getting humans to drink beverages sweetened with fructose and comparing the effect to what happens when the same people or others drink beverages sweetened with glucose — aren't applicable to real human diets, because neither humans nor animals ever naturally drink pure fructose or even pure glucose, at least not in liquid form. We always take it as pretty close to a fifty-fifty combination of the two, as in sugar and high-fructose corn syrup. And the amount of fructose or sucrose being fed to the rodents or the human subjects in these studies has typically, although not always, been enormous — usually constituting 60 or more percent of the calories in the rodents' diet, and the equivalent of 30 to 40 percent of calories from sugar in humans. What's more, these studies are short — a few months at most — and it's unclear how to extrapolate from what happens in just a few months when we're talking about conditions — metabolic syndrome, obesity, diabetes, heart disease — that develop over years and, more likely, decades. Researchers assume that it's a fair assumption that what happens in a few months on large doses of sugar (in studies that are practical and affordable) will happen over a longer period when the doses of sugar consumed are more realistic (in studies that aren't). It's a reasonable assumption, maybe a good one (I think so), but that doesn't mean it's true.

Ultimately, what the sugar industry (and researchers, both on and off the industry's payroll) will argue is that restricting sugars in these studies only decreases insulin resistance and metabolic syndrome when the subjects lose weight. They then assume that the only way to induce weight loss is to get people to eat less — a calorie is a calorie, after all, by this thinking — and so the worst that can be said about sugar is that it tastes so good, it makes people consume too many calories. This leads back to the assertion that if these people had merely eaten less or exercised more, they'd have seen similar beneficial results.

But if sugar actually causes insulin resistance — as the biochemistry and the animal experiments suggest — then it also is the very likely trigger of excess fat accumulation and thus obesity. Remove the sugar, and the insulin resistance improves and weight is lost, not because the subjects ate less, which they may have, but because their insulin resistance resolved. The sugar industry doesn't see it this way.

The attendant complexity explains why research reviews on the subject — not to be confused with the reviews by the USDA or other government agencies — typically conclude that more research is necessary. In 1993, just seven years after the FDA appeared to exonerate sugar in its report, the *American Journal of Clinical Nutrition* dedicated an entire issue to the effects of consuming fructose and thus sugar. Article after article discussed the evidence that sugar consumption might be harmful and then the need for research that did what the sugar industry's scientist-consultants had suggested two decades earlier was necessary: establish at what level of consumption sugar does, indeed, become dangerous. "Further studies are clearly needed to determine the metabolic alteration that may take place during chronic fructose or sucrose feeding," as Tappy and his colleague Éric Jéquier wrote in their review article in the special issue.

In 2010, when Tappy and his colleague Kim-Anne Lê co-authored a review on sugar, they were still reiterating the same point: "There is clearly a need for intervention studies," as they put it in the technical jargon, "in which the fructose intake of high fructose consumers is reduced to better delineate the possible pathogenic role of fructose. At present, short-term intervention studies however suggest that a high-fructose intake consisting of soft drinks, sweetened juices, or bakery products can increase the risk of metabolic and cardiovascular diseases." In less technical jargon, what's still needed is experiments that can tell us with reasonable certainty at what level or dose sugar consumption does to us what it does to laboratory rats and even baboons. Is that a higher dose than we already consume? Do we get metabolic syndrome and become insulin-resistant and so maybe obese, diabetic, and atherosclerotic because we've passed this point, or is there something else entirely to blame?

We're unlikely to learn anything more definitive in the near future, which brings us back to the issue we were discussing at the beginning of this chapter — the requirements of public-health action versus the requirements of good science. Sugar and high-fructose corn syrup are not "acute toxins," of the kind the FDA typically regulates, and the effects of which can be studied reasonably well over the course of days or months. The question is whether they're chronic toxins, their effects accumulating over the course of many thousands of meals, not just a few. This means that what Tappy referred to as "intervention studies" have to go on for years or decades to be meaningful. Thousands if not tens of thousands of subjects have to be randomized to high- and low-sugar diets and then followed for years (the more subjects in the study, the shorter the trial needs to run) to see which group experiences the greater toll in sickness and death. Such studies are exorbitantly expensive, and few researchers in this field think they'll ever be conducted.

The number of researchers interested in studying sugar and fructose and worrying about the metabolic effects of consuming them is certainly growing, as is the willingness of health organizations worldwide to fund laboratory research, or at least to discuss such funding. But this has yet to be accompanied by the kind of human trials that might identify what happens when we consume sugar or high-fructose corn syrup for years, and at what level of consumption we incur a problem. As of the fall of 2016, fewer than a dozen clinical trials — all small and of short duration — were ongoing in the United States that might actually establish anything that the researchers who pay attention to the literature haven't known for decades.

So the answer to the question of whether sugar, in the form of sucrose and HFCS, is the primary cause of insulin resistance and metabolic syndrome and therefore obesity, diabetes, and heart disease is: it certainly could be. The biological mechanisms that were elucidated by the 1970s make it clear that sugar is a prime suspect and should have been all along. The damage that these sugars do, their toxicity, would take years to accumulate and manifest themselves as disease. This wouldn't necessarily happen to everyone who ingested them (just as cigarette smoking doesn't cause lung cancer in everyone), but the biology suggests that when insulin resistance and metabolic syndrome appear, these sugars are the likely cause. The greater leap of faith, in this case, would be to assume that the sugars are harmless. And if sugars cause insulin resistance, as the evidence suggests, there are all-too-regrettable implications.

CHAPTER 10
THE IF/THEN PROBLEM: I

It is sometimes disheartening to consider that with all our abilities to detect diabetes and begin early intervention, we (i.e., IHS [the Indian Health Service] and NIH) failed to prevent the disaster that has overtaken the Tohono O'odham people and other American Indian Tribes in the United States.

JAMES W. JUSTICE,
"The History of Diabetes Mellitus in the Desert People," 1994

In February 1940, Elliott Joslin traveled to Arizona to conduct a comprehensive survey on the prevalence of diabetes in the state. He had been motivated, he would later explain, by a recent national survey that had documented large state-to-state disparities in the death rate from diabetes. Why did the states with the highest diabetes mortality — Rhode Island and Massachusetts — have a rate three to four times that of those with the lowest, of which Arizona seemed best suited for a study? Joslin was a fan of fieldwork, not "armchair statistical" research, so he took himself off to Arizona to answer the question personally. He would be aided by the state's Board of Health and its Medical Society, the Veterans' Bureau, and the Indian Health Service, all working to assure that all the red tape was cut. The local press gave his visit the necessary advance publicity, and the Phoenix Pathological Laboratory reduced its fees to a minimum for any blood-sugar tests that would have to be done. Airmail letters were sent to each of the more than 560 physicians working in the state, asking them to report back on every diabetic patient under their care.

Joslin presented his results that June at the annual meeting of the American Medical Association. His "canvass for diabetes," as he called it, had identified 755 cases in the state. Seventy-three were among the Native Americans living on reservations. After he accounted for the relative youth of the population and estimates of what percentage of cases might actually have been seen by the state's physicians, Joslin concluded that diabetes among the Native Americans in Arizona seemed no less common than it was among other ethnic groups and that the rate, in turn, was comparable to that of any other state — perhaps three or four in every thousand suffered. Diabetes, in other words, was still a rare disease at the beginning of the Second World War, both in Arizona and elsewhere, in the Native American population and among whites, but it was a universal disease. No population was exempt.

Times have changed. The prevalence of diabetes in the United States, as noted earlier, is now closer to one in eleven Americans than to the three or four in a thousand that it appeared to be when Joslin went to Arizona. As for the Native Americans in that state, by the 1960s researchers were reporting a prevalence of type 2 diabetes in adults surpassing 50 percent, the highest rate then (and perhaps since) recorded in the world. Both NIH researchers and the local physicians working for the Indian Health Service described this epidemic of diabetes as taking them by surprise. One moment the Native American population seemed to be relatively healthy, as Joslin and others had documented; if they had diabetes, the symptoms were sufficiently benign that they had no reason to be hospitalized and remained undiagnosed by the local physicians. The next moment, or so it seemed, these Native Americans were overwhelmed by the disease, as were the physicians and hospitals dedicated to providing their health care.

Understanding what happened to this Native American population is critical to understanding what's now happening to populations worldwide. How do we explain increases in prevalence of the disease of 900 percent, for instance, in the United States between the 1960s and today, if we believe the CDC statistics to be accurate? The key observations among Native American populations evolved coincident with the understanding of metabolic syndrome and insulin resistance that emerged from the 1960s onward, and so the implications are directly relevant to sugar itself and the proposition that sugar consumption is the cause.

Of the Native American tribes that have experienced diabetes epidemics, three in Arizona provide a window into what happened — the Pima (also known as the Akimel O'odham, or River People), who live along the Gila and Salt rivers, in the south-central part of the state; the Papago, a related tribe (the Tohono O'odham, or

Desert People) living farther south, and the Navajo to the northwest.

The Pima are among the best-studied indigenous populations in the world. Their history, told by missionaries, soldiers, physicians, and travelers through the Pima territory prior to the twentieth century, is of an affluent and apparently healthy population whose prosperity came to an end in the 1860s. Anglos and Mexican Americans moved into the region, overhunted the local game, and diverted for their own use the Gila River water, on which the Pima depended for fishing and irrigating their crops. In the 1870s, the Pima were experiencing what they called the "years of famine," which then extended through the end of the nineteenth century and into the twentieth. "The marvel is that the starvation, despair, and dissipation that resulted did not overwhelm the tribe," wrote the Harvard anthropologist Frank Russell, who moved to Arizona in November 1901 to study the Pima, and whose seminal report on the people and their culture was published, posthumously, four years later.

The Pima, like most Native American populations, had remained destitute and isolated — "largely bypassed by the socioeconomic developments in the rest of the

United States," as NIH researchers would later write — until the Second World War, when they were drafted into the military and began the process of integration into "white society." The decade that encompassed the war constituted what one anthropologist studying Native Americans has called the "critical juncture with modernity" for the population. During the war years, some twenty-five thousand Native Americans served in the military, and forty thousand worked in war-related industries. Both men and women of the Pima tribe took to working in the factories in nearby Phoenix. Though the economic boom sustained during the war — an estimated 250 percent increase in per capita income — didn't last, the Pima continued to acculturate to Western diets and lifestyles. The war years "accelerated the detribalization process," as a 1991 history of the wartime experience of Native Americans put it: "The reservation had contained the lives of some 400,000 persons who were cut off from the rest of American society. The war unlocked the reservation and introduced thousands of Indians, voluntarily and involuntarily, to the world beyond."

Statistics on the prevalence of obesity and diabetes in the Pima and other Native

American populations pre–World War II are scarce and come mainly from hospital records and the occasional survey by anthropologists or Indian Health Service physicians. Both Frank Russell, for instance, and a physician-turned-anthropologist named Aleš Hrdlička* commented during the first years of the twentieth century on the surprising presence of obesity among the Pima, despite their extreme poverty, although almost exclusively among the older members of the tribe, and particularly the women. They "exhibit a degree of obesity," Russell wrote, "that is in striking contrast with the 'tall and sinewy' Indian conventionalized in popular thought."

The Pima were then depending as much on government rations as on their own subsistence farming to survive. Their diet, according to Hrdlička, already consisted of "everything obtainable that enters into the dietary of the white men." Russell suggested that some item of the diet was "markedly flesh-producing," but without making any speculations about what it might be.

* Hrdlička later became the first curator of physical anthropology of what is now the National Museum of Natural History, administered by the Smithsonian Institution in Washington, D.C.

Hrdlička had also weighed and measured some 250 Pima children, equally split between boys and girls, and reported that these children were lean, if not very lean (on average), by today's standards. In 1938, a University of Arizona anthropologist weighed over two hundred Papago men applying for jobs in the Works Progress Administration and recorded that they, too, were lean, with an average weight of 158 pounds. Surveys of Papago children in the early 1940s and again in 1949 made no mention of obesity, although average weights increased by twenty pounds or more in both boys and girls between the two surveys.

As for diabetes, if it was present among the Pima in the early years of the twentieth century, neither Russell nor Hrdlička had thought it worth mention. Surveys done in the 1930s of Indian Health Service hospitals on the reservations were in accord with Joslin's survey: diabetes was still apparently a rare disease among these Native Americans. The Indian Health Service recorded just eleven deaths attributed to the disease among the entire Native American population of the state in the six years leading up to Joslin's arrival. Sage Memorial Hospital on the Navajo Reservation, a private institution, reported just a single case of diabetes between 1931 and 1936 (although, as Joslin pointed out, only seventy-five of the patients were past the age of fifty). As late as 1947, a survey of the inpatient records of twenty-five thousand Navajo admitted to the same hospital produced a total of only five cases in sixteen years.

By the early 1950s, though, evidence of the epidemic was beginning to appear. A University of Arizona survey of the health of the local Native American tribes suggested that diabetes mortality was two to three times higher than what Joslin had reported in 1940. The anthropologists carrying out the survey also noted that Pima children, despite still living in "widespread poverty," now seemed particularly prone to obesity, and that it was evident in some by age six and more often by age eleven. "That this obesity is not merely a childhood trait that is lost with physical maturity," they wrote, "is apparent to anyone who has lived or worked on the Pima Reservation for even a short period of time." A two-year survey of inpatient records in the hospitals serving the Native American population identified ninety-four cases of diabetes in the Pima, just a dozen years after Joslin had identified only twenty-one. In 1954–55, two Indian Health Service physicians, John Parks and

Eleanor Waskow, surveyed physicians and the Indian Health Service hospitals and identified 283 cases among the Pima; by their estimation, at least one in every twenty-five Pima was clearly diabetic, manifesting symptoms of the disease when it is uncontrolled.

The extent of the epidemic and the speed with which it arrived become all too clear in 1963, when two NIH researchers — Peter Bennett, a British rheumatologist, and Tom Burch, an infectious-disease epidemiologist — visited the Gila River Reservation to study rheumatoid arthritis, a disease they believed might be rare among populations like the Pima, living in hot, dry environments. Bennett and Burch took blood samples from over nine hundred Pima and found diabetic levels of blood sugar in 30 percent of them. Among those older than thirty, one in every two appeared to be an undiagnosed and untreated diabetic. Within months of reporting the results of the survey in 1965, the two NIH researchers had been reassigned to Arizona to study diabetes in the Pima and to create an NIH outpost in the state that continues to study diabetes in the Native Americans to this day. By 1971, Bennett, Burch, and their colleagues were confirming, using "conservative criteria," the highest rates of diabetes ever recorded in a population, while also noting that two-thirds of the Pima men and over 90 percent of the women were at least overweight, if not obese. Indian Health Service physicians studying the Papago and other local tribes were now beginning to report numbers almost as high.

By the mid-1980s, the epidemic of diabetes and obesity that had beset the Pima was clearly documented in the Navajo and other Native American tribes throughout Arizona, Utah, and New Mexico. Diabetes had become a primary cause of death among these populations; outpatient visits for diabetes in the Indian Health Service hospitals in Arizona nearly tripled in just a dozen years. Researchers and physicians were documenting ever-increasing levels of childhood obesity and of type 2 diabetes appearing at ever younger ages.

Throughout these decades, the Indian Health Service physicians and the NIH researchers struggled to explain what they were witnessing. How could one in two Pima adults have the blood-sugar level of a diabetic without the hospitals being full of Pima with diabetic complications? One possibility was that these Native Americans could tolerate higher levels of blood sugars than other ethnic groups, and so diabetes in these populations was a relatively benign disease. That belief was dispelled, however, as the familiar complications of diabetes — kidney disease, heart disease, hypertension, nerve damage, gangrene leading to amputation, blindness — began to appear. One NIH researcher who arrived in Arizona in 1983 to study the Pima later said he was "shocked" by "the amount of suffering" he was seeing.

The only explanation that seemed to fit, as Parks and Waskow had first suggested when they published the results of their assessment in 1961 (and as Bennett and Burch did a decade later), was that they were witnessing a wave of diabetes overtaking this population — a new disease, in effect. The Arizona hospitals hadn't been full of Native American patients with diabetic complications because these people hadn't had diabetes long enough to manifest those complications. "As more thorough examinations were done," wrote James Justice of the Indian Health Service when he reviewed the evidence in 1993, "and the duration of diabetes (mostly uncontrolled) increased, all the usual dreaded complications eventually ensued."

In 1965, when Bennett and Burch moved

permanently to Arizona to begin the study of diabetes in the Pima tribe, they were motivated by what Bennett later called, with all due respect for the tragedy unfolding, a "fantastic opportunity to try to understand diabetes itself and its implications." Over the next thirty years, the NIH researchers would learn a tremendous amount about why and how diabetes and obesity could explode in a population, as it did throughout these Native American peoples, and as it does now throughout the world.

Three factors appear to be at work.

One is the change in diet and lifestyle that these populations experienced with Westernization, which would be mirrored by aboriginal populations worldwide. By the 1980s, the NIH researchers were following the script dictated by the FDA and the NIH itself, and assuming (as Joslin and diabetes researchers had been doing since the 1920s) that the diabetes they were seeing in this Native American population was caused by the obesity that went with it. The obesity itself, they believed, was caused by an increase in calories consumed — particularly, of course, the dense calories of dietary fat — and by the sedentary behavior that these researchers assumed had arrived with more modern lifestyles. (That many of these

Native Americans were hardworking laborers and, indeed, always had been, was the kind of observation that wasn't considered meaningful in this context.)

Sugar seemed to be a prime suspect, and that was a recurring theme in a century's worth of observations and discussion. When Hrdlička had commented that the Pima were already eating Western foods in 1906, he had been referring largely to sugar, white flour, and lard purchased at local trading posts or included in the government rations. When Indian Health Service physicians studied the living conditions on the Pima, Papago, and Navajo reservations half a century later, they reported purchases of Western foods — *particularly* sugar and sweets — similar to what rural Americans elsewhere would have been purchasing from country stores thirty to forty years earlier; inevitably, the physicians also commented on the sugar in the coffee at every meal, and the "large amount of soft drinks of all types" consumed between meals. By the late 1950s, the USDA had initiated a surplus-commodity food program in which, James Justice would later report, "large quantities of refined flour, sugar, and canned fruits high in sugar" became available on the reservations. And when a physician-epidemiologist working for the CDC in 1992 wrote an essay on the explosion of diabetes now apparent in the Navajo and throughout other Native American populations, this was a point he made as well. "Even though evidence currently favors dietary fats over carbohydrates as a cause of obesity," he wrote, "the level of consumption of sugared pop by Navajo adolescents (more than twice the national average) is remarkable," and so the Indian Health Service had justifiably set program objectives to reduce both "obesity and sugared soda pop consumption."

One obvious possible explanation for the epidemics of obesity and diabetes in these Native Americans, and thus elsewhere, is that as the amount of sugar consumed per capita increases, and perhaps sugary beverages particularly, a greater proportion of the population becomes insulin-resistant. They pass over the threshold at which they can no longer tolerate the sugar they're consuming — some of us can only tolerate a little sugar; some of us can tolerate a lot — and they manifest metabolic syndrome and then obesity and diabetes. The more children eat sugar — especially as it becomes a staple of their diet in breakfast cereals, candies, ice cream, juices, and sodas — the more likely they are to manifest these problems at young ages. And if there's a lag time involved, as the South African diabetologist George Campbell had suggested in the 1960s, as there is with cigarettes and lung cancer — say, twenty years to develop diabetes after passing over the threshold — then we may still be seeing the accumulating effects in adults of those who passed over their sugar threshold decades earlier.

Genetics are also assuredly involved. Parents influence their children's likelihood to become obese and/or diabetic, not just through how and what they feed them or allow them to eat — whether and to what extent, as I'm arguing, they "ration their children's sweets" — but through their genes as well. Some of us have been passed genes that predispose us to get fat and/or diabetic in the world in which we now live, or to get fat and diabetic at younger ages than others, and these are the genes we pass on to our children. Geneticists would say some of us have susceptible "genotypes" that respond to our environment — sugar-rich, as I'm suggesting — and this is why we manifest the obese and diabetic phenotype, or manifest it at younger ages than others. Some of us don't.

Researchers studying the Pima and other

Native American tribes have assumed that their genes, for whatever reason, make them particularly susceptible to diabetes and obesity when they eat modern Western diets and live modern Western lifestyles. This may be true, but we now know that vastly different populations with (presumably) vastly different genetic inheritances suffer very similar epidemics of obesity and diabetes when their diets and lifestyles are so quickly Westernized. This suggests an alternative hypothesis, which is that all these populations — the Pima and other Native Americans — are simply the ones, as Peter Cleave suggested in the 1960s about other indigenous peoples, who had the least time to adapt to twentieth-century sugar consumption. For this reason, they were least able to tolerate its effects. They didn't have time to adapt from generation to generation, as sugar consumption slowly rose and the maladaptive nature of diabetes and obesity — birth defects and increased infant and maternal mortality — more slowly worked to create a population more in synch with its environment. Prior to the discovery of insulin, half of all diabetic mothers died during pregnancy or shortly thereafter — Joslin described the prognosis for the mother as "horrible" — and barely more than half of the fetuses or newborns survived. Other than at Joslin's clinic in Boston, the prognosis for either mother or child had barely improved, if at all, by the 1940s, even with insulin.

When clinicians and researchers in Arizona first started studying diabetes in the Pima, they assumed that if the children of diabetic mothers survived the childbirth period, "they would then be fine," as David Pettitt, a pediatrician who worked first with the Indian Health Service and then the NIH, has said. But they weren't fine. And this is where the implications are particularly dire, another possible explanation for why we are likely to be facing grave new problems moving forward if our sugar use isn't dramatically curbed.

Since 1965, with the arrival of Bennett and Burch in Arizona, the NIH has been conducting an ongoing study of diabetes in the population: Pima over the age of five have been examined every two years and followed into adulthood. As Pima women gave birth, their children were added to the study. The NIH researchers wanted to document how the wave of diabetes that had overwhelmed the Pima by the 1960s then influenced the generations that came after.

In 1983, the NIH researchers reported that *more than half* of the children who had been born to diabetic mothers had become obese by their late teens. This was more than twice the rate of obesity in children born to mothers who became diabetic only after the pregnancy, and more than three times higher than the rate for children whose mothers had been healthy throughout their pregnancy and had yet to become diabetic. In 1988, with five years more to follow these children into adulthood, the NIH researchers reported that 45 percent of the children of diabetic mothers had become diabetic themselves by the time they were in their mid-twenties, more than five times the rate among children of mothers who would go on to become diabetic only after their pregnancy (8.6 percent), and more than thirty times the rate among children of mothers who remained healthy (1.4 percent).

Clearly, genetics seemed to play a role, the NIH researchers reported, because having a father who was diabetic also increased the risk of becoming obese and diabetic early in life. But the effect of being born to a diabetic mother dwarfed that of being born to a diabetic father. This suggested that the consequences of having high blood sugar — of being insulin-resistant and thus glucose-intolerant, of having metabolic syndrome — while pregnant are passed from mother to child in the womb.

Today this concept is known as "perinatal metabolic programming" or "metabolic imprinting." The conditions in the womb — in the intrauterine environment — influence the development of the fetus, so that subtly different conditions will lead, in effect, to the birth of newborns who respond differently to the environment they face outside the womb. In particular, the nutrients that the developing child receives in the womb — including the supply of glucose — pass across the placenta in proportion to the nutrient concentration in the mother's circulation. The higher the mother's blood sugar, the greater the supply of glucose to the fetus. The developing pancreas responds by overproducing insulin-secreting cells. "The baby is not diabetic," says Boyd Metzger, who studies diabetes and pregnancy at Northwestern University, "but the insulin-producing cells in the pancreas are stimulated to function and grow in size and number by the environment they're in. So they start overfunctioning. That in turn leads to a baby laying down more fat, which is why the baby of a diabetic mother is typified by being a fat baby."

This phenomenon was first proposed by the Danish pediatrician Jorge Pedersen in the 1920s (in his doctoral thesis) and had been invoked over the intervening decades to explain why diabetic and obese mothers were more likely to give birth to very large babies. The NIH research on the Pima is just one of many studies that have now confirmed the influence of high blood sugar in pregnant women across the lifespan of their children. Women who are glucose-intolerant during their pregnancies will have children who are born larger and fatter than women who aren't, and those children will carry a greater risk of obesity and diabetes as they themselves reach adulthood. This includes not just women who are diabetic before pregnancy or become diabetic during pregnancy — a condition known as gestational diabetes — but obese women or women who gain a lot of weight in pregnancy. All these women will have higher blood sugar on average than women who remain lean and healthy; their triglycerides will be higher as well. This would explain why maternal obesity, as has been documented repeatedly, is a strong risk factor for childhood obesity and among the strongest predictors of metabolic syndrome and obesity in adulthood.

This implies, of course, that if insulin-resistant, obese, and/or diabetic mothers give birth to children who are more predisposed to being insulin-resistant, obese, and diabetic when they, in turn, are of childbearing age, the problem will get worse with each successive generation — a "vicious cycle," as it's often described in the medical literature by researchers who pay attention to the issue. It is a likely explanation for why obesity and diabetes seemed to explode in Native American populations over the course of just one or two generations, and why efforts to stem these epidemics have failed. Each successive generation includes more and more children predisposed — preprogrammed, in effect — to become obese and diabetic adults and obese and diabetic mothers. The "vicious cycle" of the "diabetic intrauterine environment," wrote the NIH research team studying the Pima in 2000, could account for much of the post–World War II increase in type 2 diabetes among this population. It might also "be a factor," they wrote, "in the alarming rise of this disease nationally." Other researchers have made the same point about the alarming rise of diabetes internationally: this vicious cycle may be driving it.

The vital question is: What initially triggers insulin resistance and metabolic syndrome and thus diabetes and obesity in all these populations — including the Pima and other indigenous populations, in which diabetes exploded through the populations over the course of a few generations, and those in which the prevalence has been increasing steadily over the course of half a century or more?

Those who hold to the conventional thinking, as we've seen, seem to bend over backward to exonerate sugar, despite the continuing accumulation of research implicating sugar as a cause, if not *the* primary cause, of insulin resistance. Because of the association of obesity and type 2 diabetes, public-health authorities and organizations such as the American Diabetes Association counsel that the key to avoiding diabetes is maintaining a healthy weight and "eating healthy." This means, as the diabetologist Frederick Allen wrote a century ago, that the "general attitude of the medical profession" to the question of whether sugar plays a causal role in diabetes "is doubtful or negative as regards statements in words. . . . But the practice of the medical profession is wholly affirmative." The ADA, for instance, calls it a "myth" that sugar causes type 2 diabetes, because that's caused by "genetics and lifestyle factors" that make us fat — i.e., "calories from any source." It then proceeds to recommend that we all avoid sugar-sweetened beverages to prevent diabetes, adding that we can "save money" by doing so. The organization accepts the role of fat accumulation in the liver as quite possibly a causal factor in the development of insulin resistance, diabetes, and obesity, but ignores the evidence building steadily since the 1980s that implicates sugars as the cause of that hepatic fat accumulation.

If sugar does cause insulin resistance, as the evidence suggests, then once populations begin to consume a sufficient amount — whatever that amount might be — and once the women in these populations begin to manifest metabolic syndrome, once they begin to get fatter and insulin-resistant, once this insulin resistance and glucose intolerance manifest themselves during pregnancy, then the epidemics of obesity and diabetes may be preordained. They may happen quickly, as they have in indigenous populations exposed over the course of a few decades to the sugar-rich environment of twentieth-century Western populations, or they may happen more slowly. But they will happen. And as the NIH researchers wrote in 1988 when discussing this problem in the Pima, there may be no going back. "It is unknown," they wrote, "whether this cycle can be broken." Treating diabetes and high blood sugar during pregnancy is obviously one way to do so, and physicians now work hard to do just that. Identifying the ultimate cause of the insulin resistance, though, even acknowledging the possibility that it could be sugar, would have far more profound consequences.

Chapter 11
The If/Then Problem: II

PROVISIONAL LIST OF WESTERN DISEASES

Metabolic and cardiovascular: essential hypertension, obesity, diabetes mellitus (type II), cholesterol gallstones, cerebrovascular disease, peripheral vascular disease, coronary heart disease, varicose veins, deep vein thrombosis, and pulmonary embolism

Colonic: constipation, appendicitis, diverticular disease, haemorrhoids; cancer and polyp of large bowel

Other diseases: dental caries, renal stone, hyperuricaemia and gout, thyroidtoxicosis, pernicious anaemia, subacute combined degeneration, also other forms of cancer such as breast and lung

HUGH TROWELL AND DENIS BURKITT,
Western Diseases: Their Emergence and Prevention, 1981

In 1981, when Hugh Trowell and Denis Burkitt published their provisional list of Western diseases, there was little controversy about it, and there still isn't. Western diseases were mostly chronic disorders, not infectious diseases, and they associated with Western diets and lifestyles, common in Europe and the United States and in urban centers elsewhere, and relatively uncommon in indigenous populations isolated from Western influence. Despite the presence of such diseases as breast and colon cancer on the list, the implication of this clustering of diseases with Westernization is that they are caused not necessarily by industrial chemicals in the environment or by bad luck, but by something in the food we now eat or the way we live.

Both Trowell and Burkitt had begun their careers as missionary physicians. Trowell had spent thirty years working and teaching in the hospitals and medical schools of Kenya and Uganda. In 1960, the year after his retirement, he had published *Non-Infectious Diseases in Africa,* a book that represented the first concerted effort to document the spectrum of diseases afflicting the native population of the continent. Burkitt had worked for eighteen years in Uganda and had become, in the process, what *The Washington Post* would later call "one of the world's best-known medical detectives." This praise was for Burkitt's pioneering epidemiological studies, leading to the identification of the first human cancer ever linked to a viral cause, a fatal childhood malignancy known since as Burkitt's lymphoma.

Burkitt and Trowell based their provisional list of Western diseases on their surveys of hospital inpatient records worldwide, on the existing medical literature, and on the suggestions of the thirty-four physician-researchers from five continents who contributed to the book *Western Diseases: Their Emergence and Prevention.* They called it a "provisional list" because they acknowledged that such a pioneering effort was likely to contain errors, and because other diseases already appeared likely to be added to it — including irritable bowel syndrome, ulcerative colitis, Crohn's disease, and autoimmune disorders — but the evidence for those potential additions was not yet sufficient. The list was a much-expanded version of the diseases that Peter Cleave and George Campbell had called "saccharine diseases" in the 1950s, implying that refined grains and sugars were to blame (Burkitt and Trowell credited Cleave with being a guiding light in their work), and that Yudkin was discussing and referring to in 1963 as "diseases of civilization," which was the more commonly used term at the time.

Trowell and Burkitt preferred to call them "Western diseases" for what in retrospect was an obvious reason: "It proved obnoxious," they wrote, "to teach African and Asian medical students that their communities had a low incidence of these diseases because they were uncivilized." It's their terminology that's still with us today. These diseases have tended to increase in prevalence through the twentieth century and into the twenty-first, and many of them are closely associated with obesity and type 2 diabetes.

We can think of Burkitt and Trowell's provisional 1981 list as a product of the collective medical consciousness of the British Empire. One of the advantages of having colonies, protectorates, dominions, and territories scattered over much of the planet is that it allows for the physicians working in these far-flung locales — "where the conditions of life differ so widely," Joseph Chamberlain, colonial secretary (and father of Neville), would phrase it in 1903 with the founding of the British Cancer Research Fund — to compare and contrast their clinical experiences and inpatient records with those of their colleagues working in the home country. Physicians like Burkitt and Trowell had the opportunity to train in British medical schools and hospitals and then ply their trade in missionary or colonial hospitals in far-off corners of the empire. They could see firsthand the differences in the spectrum of diseases afflicting Europeans and the indigenous populations to which they administered — differences in the "pattern and pathogenesis of disease," as one such physician, John Higginson, founding director in 1965 of the International Agency for Cancer Research, would later describe this observation. And they could also observe how the disease spectrum of these indigenous peoples changed with time as they adapted to Western diets and urban lives.

When Trowell arrived in Kenya in 1929, for instance, the region already had a local medical association with a professional journal — the *East African Medical Journal,* founded in 1923 — and well over a hundred physician-members, all, like Trowell, trained and qualified in Europe. Their job was to see to the health of the thousands of British settlers who had begun moving into the region, and to the three million native

Africans already there and still largely living as they had been for untold generations. "Never before," Trowell wrote, "and probably never again will . . . so many resident doctors observe three million men, women and children, as in Kenya in the 1920s, emerge from preindustrial tribal life and undergo rapid westernization."

What Trowell and his colleagues experienced in Kenya and Uganda, though, was only a variation on George Campbell's observations in South Africa, the findings of the Indian Health Service physicians working on reservations in Arizona and throughout the United States, and the information gathered by all those physicians and researchers who documented the arrival of diabetes in indigenous populations worldwide.

When Trowell arrived in Kenya, he would later write, hypertension and diabetes were absent. The native population was also as thin as "ancient Egyptians," despite consuming relatively high-fat diets and suffering no shortage of food.* By the 1950s,

* During World War II, according to Trowell, the British government sent a team of nutritionists to the region to learn why local Africans recruited into the British Army could not gain sufficient

obese Africans were a common sight in the towns and cities. In 1956, Trowell himself reported what he believed to be the first diagnosis of coronary heart disease in a black African, an obese High Court judge who had spent two decades living (and thus eating) in England. By the 1960s, hypertension was as common among black Africans as it was in any other population in the Western world. When Trowell returned to East Africa in 1970, "the towns were full of obese Africans and there was a large diabetic clinic in every city. The twin diseases had been born about the same time and are now growing together."

Burkitt and Trowell observed, as Cleave, Campbell, and Yudkin had observed before them, a consistent pattern of pathogenesis in the British medical literature and in the observations of hundreds if not thousands of physicians worldwide. When populations underwent Westernization, chronic diseases

---

weight to meet army entrance requirements. "Hundreds of x-rays," Trowell wrote, "were taken of African intestines in an effort to solve the mystery that lay in the fact that everyone knew how to fatten a chicken for the pot, but no one knew how to make Africans . . . put on flesh and fat for battle. It remained a mystery."

emerged with it, whether rapid or not, and typically in the same order, beginning with periodontal disease (tooth decay), gout, obesity, diabetes, and hypertension, and eventually encompassing all of them.

Because this pattern of pathogenesis differs from population to population in its details and specifics, to understand exactly what is happening, and perhaps why, requires the perspective of evolutionary biology. "The incidence and variety of diseases in a community reflects always the interplay of many environmental factors on the genetic pool of the community," wrote Burkitt and Trowell in their preface to *Western Diseases.* The genes or genotype of any two populations will differ, as will the genes of the individuals in those populations, although to a lesser extent. The environment in which those genes manifest themselves and have for generations will also differ. This means that the influence of Westernization will have a different impact on each population and each individual, but the general patterns will be the same. "In relatively stable populations," wrote Burkitt, "the community genetic pool alters only very slowly during long periods of evolutionary time; in comparison the environment may alter very quickly. If environmental factors change *rapidly* then the pattern of environment-related diseases also changes rapidly."

It seemed a very good bet, Burkitt argued, that if a cluster of associated diseases appeared at the same time in a population or worldwide, those diseases had a common cause. This was the simplest possible hypothesis. In 1975, when Burkitt discussed what he called the "significance of relationships" in the first book he and Trowell had co-edited on these Western diseases, he pointed out that a single environmental trigger could result in a wide spectrum of diseases depending on the genetic variation in the individuals exposed, the duration of exposure, and the amount of exposure over time and in individuals.

One of Burkitt's examples was cigarettes. The first symptom of smoking was likely to be stained fingers (back in the days of mostly unfiltered cigarettes), often to be followed by bronchitis and eventually lung cancer. Had he known at the time, Burkitt might have added emphysema and heart disease. The appearance of these disorders in individuals would depend on how long they smoked and how much they smoked, and on their individual susceptibility. Some lucky individuals or those genetically blessed would seem immune to all these conditions, and would get nothing more than stained fingers, despite smoking packs a day. Some would get bronchitis, some bronchitis and lung cancer, some only lung cancer. Not every individual would get every manifestation of this disease pattern, but all the smoking-related diseases would appear in the population, and smoking cigarettes would be the cause of all of them. Only by comparing populations with and without cigarettes — or smokers to nonsmokers within a population — would researchers be able to clarify the patterns and the causality.

Syphilis was another example. "Before the spirochaete of syphilis had been identified," Burkitt wrote, "the association in individual patients of several manifestations of this disease must have suggested a common cause. Palate perforation, sub-periosteal bone deposits and a previous history of a characteristic skin rash and penile sore would have been observed in a single patient." If untreated, it would eventually manifest itself in dementia, deafness, and heart and nerve damage, yet all caused by the same, single agent. "If this characteristic pattern of emergence of certain diseases occurs in communities previously almost exempt from these disorders," Burkitt continued, with "early," "mid," and "late" arrival conditions determined by the duration of exposure, "this suggests a common causative factor or associated causative factors."

In Burkitt and Trowell's provisional list of diseases caused by exposure to a Western lifestyle, conditions such as appendicitis and tooth decay appeared typically in childhood. These didn't require a long-lived population to manifest themselves, and should appear earliest after the transition to Westernization. This would make it relatively easy to identify their cause. Obesity, diabetes, gout, and hypertension, among other diseases, tended to appear only as individuals in the exposed population passed into middle age. Cancers and heart disease might typically require an exposure of fifty or more years before they appeared, and thus represented a particular challenge: the indigenous populations being served by these missionary and colonial physicians tended to be relatively short-lived, so a relative absence of a disease like cancer could in reality be a relative absence of individuals in the population old enough to get cancer or seek treatment for it.

In Cleave's books on what he called the

saccharine disease, he had suggested that tooth decay provided the obvious clue to the causality of this clustering of Western diseases. Appearing early in life, he said, it was the equivalent of the canary in the coal mine and foretold the coming of the entire spectrum of Western disease. Since tooth decay was caused by refined grains and perhaps sugar most of all, Cleave argued, didn't that imply that the same would be true of all these Western diseases? "It would be an extraordinary coincidence," he wrote, "if these refined carbohydrates, which are known to wreak such havoc on the teeth, did not also have profound repercussions on other parts of the alimentary canal during their passage along it, and on other parts of the body after absorption from the canal."

In 1975, when Burkitt and Trowell published their first book on these Western diseases, they were thinking the same way, although their preferred explanation was that it was the absence of fiber in modern processed foods that was primarily responsible. Fiber was removed in the processing of sugar and grains, and constipation was also an "early" disorder in the cluster, the one (and perhaps only) disorder that appears to be treated or prevented by the addition of fiber to a diet.

By 1981, when they published *Western Diseases,* Burkitt and Trowell had embraced a more conventional view of the problem. Nutrition researchers in the 1970s had focused their attention almost exclusively on saturated fat as the cause of heart disease and salt as the cause of hypertension. Burkitt and Trowell went along with their peers and adopted a less parsimonious way of viewing the emergence of these Western diseases.

But is this perspective justified? Can a host of chronic diseases that cluster together both in individuals and in populations and associate closely with Western diets and lifestyles best be explained by the presence of a single dietary trigger — i.e., sugar — or by multiple triggers? When Isaac Newton paraphrased the concept of Occam's Razor, he did so by saying, "We are to admit no more causes of natural things than such as are both true and sufficient to explain their appearances." This was rule number one of Newton's "rules of reasoning in natural philosophy" in his *Principia.* So is it necessary to posit multiple aspects of diet and lifestyle — multiple causes — to explain the presence of these chronic diseases that associate with Western and urban lives, or will one suffice? Sugar, for example.

Consider, for instance, the relationship between obesity, diabetes, heart disease, and gout. The latter three are associated with obesity, and the conventional thinking is that they are caused by, or exacerbated by, the accumulation of excess fat — obesity. All four cluster together in populations and in individuals. All are associated as well with hypertension and considered by physicians to be hypertensive disorders, which means blood pressure tends to be pathologically elevated in all of them. This would imply that all these diseases are likely to be caused by the same dietary or lifestyle trigger, whatever it is. But by the 1980s, this was no longer how they were seen.

The single best-documented example of the clustering of these diseases and how they appear together in populations following Westernization happens to be found in studies of an island nation in the South Pacific known as Tokelau, which now has the highest prevalence of diabetes of any single nation in the world (not to be confused with any single population, such as the Pima). As of 2014, almost 38 percent of all Tokelauans had been diagnosed with diabetes. More than two-thirds were obese.

Here we have an epidemiologic snapshot of how life changed with Westernization that is unparalleled in the annals of nutrition research. Tokelau is a protectorate of New Zealand, a cluster of three atolls. In the 1960s, as the Tokelau population grew to almost two thousand islanders, the New Zealand government instituted a voluntary migration program to the New Zealand mainland. In 1968, epidemiologists led by Ian Prior of the Wellington School of Medicine launched the Tokelau Island Migrant Study (TIMS) to document the diet and health of every single Tokelauan who immigrated, following them through the relevant transition to more Western and urban lifestyles, and of all those who remained behind on the atolls.

Through the mid-1960s, as TIMS got up and running, the Tokelauans had subsisted on a diet of coconut, fish, pork (fed on coconuts and fish), chickens, a starchy melon called breadfruit, and another starchy root vegetable known as pulaka. The diet had among the highest fat concentrations in the world at the time — more than 50 percent of the calories consumed came from fat, and most of that was saturated fat from the coconuts. In 1968, the islanders were already consuming some sugar and white flour delivered by the occasional trading boat, but still little by modern Western standards — 2 percent of their total calories, which works out to an annual average of less than eight pounds of sugar per islander. The medical records of the islanders at the time documented bouts of chicken pox, measles, occasional cases of leprosy, skin diseases, and asthma — and a few had gout. Three percent of the men and almost 9 percent of the adult women were diabetic.

The change to a more Western dietary pattern occurred gradually on the atolls and then accelerated in the late 1970s with the adoption of a cash economy and the establishment of trading posts on the island. By 1982, in the last TIMS assessment, coconut consumption had decreased. Per capita sugar consumption had increased to fifty-four pounds per year, and the consumption of white flour had jumped from twelve pounds per person annually to seventy pounds. Alcohol consumption increased, and cigarette smoking became more prevalent. Tinned meats and frozen foods arrived on the islands as well, although they were eaten in relatively trivial amounts compared with the normal diet of fish.

The diet and lifestyle changes for the Tokelauans who immigrated to New Zealand were abrupt and even more dramatic. Bread and potatoes replaced breadfruit in their diets; meat replaced fish; they hardly ate any coconuts. Sugar consumption skyrocketed, as did *physical activity:* the men went to work as manual laborers in the forest service or on the railway, and the women got jobs in electrical assembly plants or clothing factories, or they cleaned offices during the evening hours, walking miles to and from work.

In both populations, a similar pattern of chronic diseases erupted with the Westernization of the diet. Between the late 1960s and early 1980s, diabetes prevalence shot upward, particularly among the immigrants. By 1982, almost 20 percent of the immigrant women and 11 percent of the immigrant men — one in five and one in nine, respectively — were diabetic. Hypertension, heart disease, and gout also increased significantly, particularly in the migrant population (the migrants were nine times as likely to get gout as those remaining behind on the atolls). Obesity, unsurprisingly, also increased: Both men and women gained, on average, between twenty and thirty pounds. Children, too, got fatter.

What's to blame?

As the Tokelau experience demonstrates, Westernization brings with it significant changes in diet and lifestyle, and thus significant challenges to establishing causality. Records of the foods and drinks delivered to Tokelau far more recently (between 2008 and 2012), as collected from the manifests of the trading vessels making regular trips, document huge amounts of white rice, sugar, and flour, of hard liquor, beer, soft drinks, cigarettes, and plenty of other modern foods as well — meats, ice cream, butter, even fruits and vegetables not native to the atolls. Any or all of it could be working to increase the occurrence of the spectrum of Western diseases.

The conventional thinking about this problem, which arose from the nutrition research in the United States in the 1960s and 1970s, is that each of the Western diseases has different dietary and lifestyle triggers, even though the conditions are part of a single cluster of related diseases. Ian Prior and his colleagues suggested that in TIMS "a different set of relevant variables might account for observed differences in [disease] incidence," but simultaneously acknowledged that the contrasting experience of the migrants and those who remained behind on Tokelau made this attribution of multiple causes surprisingly difficult to do.

The migrants gained more weight than

the atoll dwellers, even though the migrant lifestyle was significantly the more active of the two. And even though the migrants manifested increasing evidence of heart disease, their diets contained significantly *less* saturated fat than what they had been eating on Tokelau. Prior and his colleagues suggested that excess weight (eating too much) was at least partially responsible for the increases in hypertension, gout, diabetes, and heart disease among the migrants. And because the migrants seemed to eat more salt, this could also explain the increased prevalence of hypertension, as could the stress of assimilation to a new culture. The migrants ate more red meat than the atoll dwellers, which could explain why so many of them were getting gout. An increase in asthma on the mainland of New Zealand might be explained by the presence of allergens that were absent on the mainland.

All of this makes sense, and it's more or less how we still think of these diseases today. But I'm writing about sugar for a good reason: because Burkitt's logical analysis about causality is correct. The simplest hypothesis — as encapsulated in Occam's Razor — is always *the most likely.* It may not turn out to be right; the perpetrator of the first of a series of apparently related crimes in a community is not necessarily responsible for all of them, but it is the most likely hypothesis that he or she *is* responsible, and the one that should be considered and perhaps ruled out before multiple perpetrators or hypotheses are suspected. Because the kind of observational evidence researchers deal with is incapable of establishing beyond reasonable doubt that sugar (or any other dietary suspect, for that matter) *is* the factor in Western diets and lifestyles that triggers the aforementioned cluster of chronic diseases, the best we can do is ask whether this is a likely possibility, and if so, whether it is, indeed, the *most* likely.

What makes sugar the leading candidate by far (and what should have made it so when Prior and his colleagues were trying to understand what they were observing in TIMS) is the revelations about metabolic syndrome and insulin resistance. These shifted the obesity/diabetes/heart-disease paradigm from the conventional thinking of the 1970s — obesity is caused by eating too much, diabetes by being too fat, and heart disease by some combination of the two plus the saturated fat in our diets — to the current perspective, according to which metabolic syndrome is the critical player in obesity, heart disease, and diabetes. The fact that many of the Western diseases in Burkitt and Trowell's list, these chronic disorders that associate with Western diets and life-styles, are also diseases that associate with obesity and diabetes puts the focus, in turn, on insulin resistance and metabolic syndrome as a mechanism or at least a critical precursor. And if insulin resistance and metabolic syndrome are ultimately caused by the sugars we consume, then so are, to some extent, *all* these other diseases as well. *This* is why sugar should be at the top of any list of dietary suspects.

For the past fifty years, as the Tokelau case illustrates, nutritionists and heart-disease researchers have assumed that eating too much salt is the cause of hypertension, which can be defined as chronically and pathologically high levels of blood pressure. That hypertension is one of the five criteria that a physician will use in diagnosing metabolic syndrome would make it seem obvious that it's likely caused by the same trigger — dietary or otherwise — as the other conditions. In other words, if your blood pressure is elevated, that's a sign that you're insulin-resistant and have metabolic syndrome; it also means you're likely to be overweight, or at least getting fatter, and your triglycerides are elevated, you're glucose-intolerant, and your HDL cholesterol is low. They all go hand in hand and are probably caused by the same thing. By Occam's Razor and Burkitt's logic, *if* sugar causes insulin resistance and elevates triglycerides and makes us fat, *then* it very likely causes hypertension, too — if not directly, then at least indirectly, through its effect on insulin resistance and weight. Sugar is the culprit.

So here's the if/then hypothesis: *If* these Western diseases are associated with obesity, diabetes, insulin resistance, and metabolic syndrome, which many of them are, *then* whatever causes insulin resistance and metabolic syndrome is likely to be the necessary dietary trigger for the diseases, or at least a key player in the causal pathway. Because there is significant reason to believe that sugars — sucrose and high-fructose corn syrup in particular, the nearly fifty-fifty combinations of glucose and fructose — *are* the dietary trigger of insulin resistance and metabolic syndrome, it's quite likely they are a primary cause of all these Western diseases, including, as we'll discuss, cancer and Alzheimer's disease. Without these sugars in the diet, these chronic diseases would be relatively rare, if not, in some cases, virtually nonexistent.

I want to review the major Western diseases, one by one, to discuss the likelihood that sugar is responsible, or at least largely responsible — *a* prime suspect, if not *the* prime suspect. We've already discussed obesity and diabetes at length, and also heart disease, indirectly, through its relationship with insulin resistance and metabolic syndrome. So let's begin here with gout, and then we'll return to hypertension and go on to cancer and Alzheimer's disease — or senile dementia — a nightmare disorder that wasn't even on Burkitt and Trowell's radar in the 1970s and 1980s.

Gout is particularly interesting because it is clearly an ancient disease — signs of its ravages can be seen in skeletal remains, Egyptian mummies, from seven thousand years ago — and yet it's also the very first chronic disease to be indisputably linked to (relatively) modern diets and lifestyles, particularly overconsumption, however we choose to define it. Gout is rarely the subject of media attention, and yet it is more prevalent than ever. Recent surveys suggest that nearly 6 percent of all American men over the age of twenty suffer from gout, and more than 2 percent of women. The proportion rises with age, to over 9 percent of men and women in their seventies and over 12 percent in their eighties — almost one in every eight. Gout prevalence more than doubled from the 1960s to the 1990s, in association with the increases in obesity and diabetes. It appears to have increased steadily since then.

The pathology of gout has been understood since the mid-nineteenth century, when the British physician Alfred Garrod identified the compound called uric acid as the critical agent; uric acid accumulates in the circulation (hyperuricemia) to the point that it falls out of solution, as a chemist would put it, and crystallizes into needle-sharp urate crystals. These crystals then lodge in the soft tissues and in the joints of the extremities — classically, the big toe — and cause inflammation, swelling, and an excruciating pain that was described memorably by the eighteenth-century bon vivant Sydney Smith as akin to walking on one's eyeballs.

The questions then become: where does the uric acid itself come from, and why so much of it? Because uric acid itself is a breakdown product of protein compounds known as purines — building blocks, among other things, of amino acids — and because purines are at their highest dietary concentration in meat, it has been assumed for more than a century that a primary means of elevating uric acid levels in the blood, and thus causing first hyperuricemia and then gout, is an excess of meat consumption. But this is the kind of hypothesis that has been hard to confirm in experimental tests. Or, as two Harvard physicians, Friedrich Klemperer and Walter Bauer, put it elegantly in a 1947 medical textbook, "It is a most regrettable circumstance that these teachings, which are shrouded in the semisanctity of a long and venerable heritage, have never been tested by either adequate experimentation or comprehensive statistical analysis of clinical data."

As it turns out, a nearly vegetarian diet is likely to have only a very modest effect on uric acid levels — at least compared with a typical American diet — rarely sufficient to return high uric acid levels to normality, and there's little evidence that such diets reliably reduce the incidence of gouty attacks in those afflicted. This is why purine-free diets are no longer prescribed for the treatment of gout, as the physician and biochemist Irving Fox noted in 1984, "because of their ineffectiveness" and their "minor influence" on uric acid levels. The incidence of gout in vegetarians, or mostly vegetarians, has always been significant and "much higher than is generally assumed," as Bauer and Klemperer wrote, noting that one mid-century estimate put the incidence of gout in India among "largely vegetarians and teetotalers" at 7 percent. Eating *more* protein, which is, of course, found in high levels in red meat, apparently increases the excretion of uric acid from the kidneys and, by doing so, reportedly decreases the level of uric acid in the blood. This implies that the meat/gout hypothesis is very debatable; the high protein content of meats could be beneficial, even if the purines are not.

If meat isn't the cause (and those "teetotalers" suggest that alcohol alone cannot explain the presence of gout), what is?

The first clue is the association between gout and the entire spectrum of Western diseases, and between hyperuricemia and the metabolic abnormalities of insulin resistance and metabolic syndrome. In the past century, gout has manifested all of the familiar patterns, chronologically and geographically, of Western diseases. In primitive populations eating traditional diets, gout was virtually unknown or at least went mostly unreported. In 1947, Trowell reported that the disease was so rare in East Africa that he had never seen a case personally in a native African, or even read of one, in the first seventeen years of his practice. When he finally did treat a Rwandan native for gout, Trowell found it sufficiently notable that he published a case report in the *East African Medical Journal.* Even in the 1960s, hospital records from Kenya and Uganda suggested an incidence of gout lower than one in a thousand among the native Africans. By the late 1970s, however, uric acid levels in Africa were increasing with Westernization and urbanization, while the incidence of both hyperuricemia and gout among South Pacific Islanders was skyrocketing. In 1975, the New Zealand rheumatologist B. S. Rose, a colleague of Ian Prior's, described the native populations of the South Pacific as "one large gouty family."

Gout has been linked to obesity since the Hippocratic era, and this association is the origin of the assumption that high living and excessive appetites are the cause. Gouty men have long been reported to suffer higher rates of atherosclerosis and hypertension; stroke and coronary heart disease are common causes of death. Diabetes is also commonly associated with gout. In 1951,

Harvard researchers reported that serum uric acid levels rose with weight, and that men who suffered heart attacks were four times as likely to be hyperuricemic as healthy controls. This led to a series of studies in the 1960s, as clinical investigators first linked hyperuricemia to glucose intolerance and high triglycerides, and later to high insulin levels and insulin resistance. By the 1990s, Gerald Reaven at Stanford, among others, was reporting that insulin resistance and hyperinsulinemia raised uric acid levels, apparently by decreasing the excretion of uric acid by the kidney. "It appears that modulation of serum uric concentration by insulin resistance is exerted at the level of the kidney," Reaven wrote. Therefore, the more insulin-resistant an individual, the higher the serum uric acid concentration.

The evidence for sugar or fructose as a primary cause of gout is twofold.

First, the circumstantial evidence: not just the appearance of gout in isolated populations as they become Westernized and urbanized, but in Europe and America as well. The distribution of gout in these populations has paralleled the availability of sugar for centuries. Until the late seventeenth century, the disease afflicted almost exclusively the nobility, the rich, and the educated — those who could afford to indulge an excessive appetite for food and alcohol — and reached almost epidemic proportions among them in Britain. Gout then spread throughout British society in the eighteenth century. Historians refer to this as the "gout wave," and it closely parallels the birth and growth of the British sugar industry and the transformation of sugar (borrowing, once again, Sidney Mintz's phrase) from "a luxury of kings into the kingly luxury of commoners."*

The second piece of evidence is much less circumstantial: the fructose component of sugars increases serum levels of uric acid. The "striking increase" in those levels with an infusion of fructose was first reported in the late 1960s by Finnish researchers, who referred to it as "fructose-induced hyperuricemia." This was followed by a series of studies through the late 1980s confirming the existence of the effect and reporting on the variety of biochemical mechanisms by which it came about. When fructose is metabolized in the liver, for instance, it accelerates the breakdown of a molecule

* Part of this gout wave may also have been caused by lead contamination in the fortified wines — port, for instance — being consumed at the time.

called ATP, which is the primary source of energy for cellular reactions and is loaded with purines. ("ATP" stands for "adenosine triphosphate"; adenosine is a form of adenine, a purine.) This in turn increases the formation of uric acid. Alcohol raises uric acid levels through the same mechanism (although beer also has purines in it). The effect of fructose on ATP also works to stimulate the synthesis of purines, and the metabolism of fructose leads to the production of lactic acid, which reduces the excretion of uric acid by the kidney and thereby raises uric acid concentrations indirectly.

These mechanistic explanations of how fructose raises uric acid levels were then supported by a genetic connection between fructose metabolism and gout. The disease often runs in families, so much so that clinicians studying gout have always assumed the disease has a strong hereditary component. In 1990, a collaboration led by Edwin Seegmiller, a pioneer of gout research in the United States, and George Radda, who would later become director of the U.K. Medical Research Council, reported that the explanation for this familial association seemed to be a very specific defect in the genes that regulate fructose metabolism. Individuals who inherit this defect will have trouble metabolizing fructose and will thus be born with a predisposition to gout. This suggested the possibility, the researchers concluded, that the defect in fructose metabolism was "a fairly common cause of gout."

As these observations appeared in the literature, the researchers making them were reasonably clear about the implications: "Since serum-uric-acid levels are critical in individuals with gout, fructose might deserve consideration in their diet," noted the Finnish researchers in 1967; the chronic consequences of high-fructose diets on healthy individuals required further evaluation. Gouty patients should avoid high-fructose or high-sucrose diets, explained an article on nutrition and gout in 1984, because "fructose can accelerate rates of uric acid synthesis as well as lead to increased triglyceride production." In 1993, the British biochemist Peter Mayes published an article on fructose metabolism in the *American Journal of Clinical Nutrition* that reviewed the literature and concluded that high-fructose diets in healthy individuals — in other words, high-sugar diets — were likely to cause hyperuricemia and, by implication, gout as well, but the studies to address that possibility were never conducted.

This, in addition to Reaven's research reporting that high insulin levels and insulin resistance will increase uric acid levels, suggests that sucrose and high-fructose corn syrup would constitute the worst of all carbohydrates when it comes to uric acid and gout. The fructose would increase uric acid production and decrease uric acid excretion, while the glucose, through its effect on insulin, would also decrease uric acid excretion. It would be reasonable, therefore, to assume or at least to speculate that sugar is a likely cause of gout, and that the patterns of sugar consumption explain the appearance and distribution of the disease.

This hypothesis has only been seriously considered in the last few years. Those nutrition researchers interested in gout focused almost exclusively on alcohol and meat consumption. The historical belief that gouty individuals, particularly obese gouty individuals, should shy away from meat and alcohol fit in well with the dietary prescriptions of the 1970s onward.

The sugar/fructose hypothesis was ignored, once again because of bad timing. In the mid-1960s, the pharmaceutical industry developed an inexpensive drug called allopurinol that could lower uric acid levels and could be used by those with gout to prevent future attacks of the disease. The clinical investigators whose laboratories were devoted to studying the mechanisms of gout and purine metabolism began focusing their efforts either on working out the nuances of allopurinol therapy or on applying the new techniques of molecular biology to the genetics of gout and rare disorders of hyperuricemia or purine metabolism. Nutritional studies were simply not considered worthy of their time, if for no other reason than that allopurinol appeared to allow gout sufferers to eat or drink whatever they wanted.

This development coincided with the emergence of research on fructose-induced hyperuricemia. By the 1980s, when the ability of fructose and sucrose consumption to raise uric acid levels in human subjects was demonstrated repeatedly, the era of basic research on gout had come to an end. The major players had left the field and NIH funding for the study of gout had dwindled to a trickle. When the major medical journals ran occasional articles on the clinical management of gout, these would concentrate almost exclusively on drug therapy. Discussions of diet would be only a few sentences long, and typically the science in them was confused. Articles on the dietary treatment of gout — even those informed on the relationship between insulin resistance and uric acid — might include "sugars" and "sweets" as among the recommended foods with low-purine contents. In a few cases, articles that did this also noted that fructose consumption raises uric acid levels, suggesting only that the authors had been unaware of the role of fructose in "sugars" and "sweets."

Recent research on fructose-induced hyperuricemia indicates that the implications for human physiology and, in this case, pathology may extend far beyond gout itself. Since the late 1990s, Richard Johnson, a kidney specialist now at the University of Colorado, has been studying the effect of uric acid on the blood vessels leading into the kidneys. If uric acid levels in the circulation are high enough, this might damage these blood vessels and, in so doing, elevate blood pressure. And if sugar consumption is raising uric acid levels, it's a reasonable assumption that sugar consumption elevates blood pressure. This is another potentially harmful effect of fructose and sugar that was discovered only after the FDA's official 1986 exoneration of sugar in the diet (like DNA evidence implicating the prime suspect in a murder that comes along only after the suspect has been tried and acquitted for lack of evidence). It's yet another mechanism by which sucrose and high-fructose corn syrup could be a particularly unhealthy combination, and would potentially explain the common association of gout and hypertension, and even of diabetes and hypertension, although it's only one of several such mechanisms.

For fifty years, the consensus of opinion in the medical community has been that the dietary trigger of hypertension is salt consumption. Eating too much salt raises blood pressure; hypertension is the pathological, chronic state that in turn increases risk of both heart disease and cerebrovascular disease (strokes). It's a simple hypothesis and a concise one — and it's all too likely wrong. But to suggest that sugar causes hypertension is to suggest that salt doesn't (or not as much), and public-health authorities typically take umbrage. So it's necessary to talk this through, beginning with some history.

Hypertension is yet another example of how perspective and the available technology drive scientific understanding. In this case, before medical researchers could begin to understand what it meant to have high blood pressure, and who had it and who didn't, and then establish its link to other diseases, particularly heart disease and stroke, they required a relatively easy and standardized way to measure blood pressure in patients. Not until the early twentieth century was such a device, the sphygmomanometer, readily available to practicing physicians. It was the early version of the upper-arm cuff still in use today. In the 1920s, physicians around the world started measuring blood pressure in isolated, aboriginal populations so that their blood pressure could be compared to the blood pressure of those who ate modern Western diets and lived modern Western lifestyles. Physicians in the United States and Europe were debating whether high blood pressure was a bad thing or a good thing (perhaps a compensatory response of the body to nourish tissues that were having trouble getting enough blood, "a saving process in spite of the fact that it carries possibilities of harm in its possessor," as one 1920 textbook suggested). It was life-insurance actuaries, with money riding on the outcome, who first did what would become the definitive research.

By the 1920s, these actuaries had established a few unambiguous facts about blood pressure and hypertension: In particular, blood pressure increases with age *and* with weight, or at least it does in Europe and the United States (just as the likelihood of having diabetes does), and then, of course, weight itself increases with age. Among the middle-aged men a century ago who considered themselves healthy enough to apply for life insurance, systolic blood pressure below 140 millimeters of mercury (mm Hg) seemed relatively benign, which is why this number is still considered the *lower* bound of hypertension. As blood pressure went up from 140, prospects for a long and healthy life went down, and so the life insurance companies were hesitant to insure individuals with blood pressure at that level and above, or at least to insure them at the same rates as men with lower blood pressure. The insurance companies would lose money if they did — more "claims would have to be paid," as the chief medical director of the Mutual Life Insurance Company wrote in *The Journal of the American Medical Association* in 1923.

After another twenty years of study, it was clear that what was true about blood pressure in the United States and Europe wasn't the case in indigenous populations that had yet to be exposed to Western diets and lifestyles. Just as diabetes and obesity seemed rare to nonexistent in these populations, so was this characteristic increase of blood pressure with age. Blood pressure tended to be lower at young ages, and stayed resolutely low throughout life, an observation that was first reported in the Philippines and then among Zuni Indians in New Mexico, the Inuit in Greenland and Labrador, native tribes in Kenya ("This contrast" between blood pressure in the African tribes and among the local Europeans "is somewhat striking and seems to require explanation"), Bedouin tribes in Syria ("the conspicuous hypotension [low blood pressure] of the Arab"), Chinese aboriginal populations, indigenous peoples of the Yucatán and Guatemala, and, as World War II was coming to an end, among Kuna Indians in Panama ("a striking finding is the total absence of hypertension"). By the 1960s, as these populations became urbanized and Westernized, physicians — Hugh Trowell among them — were reporting that hypertension had emerged in these populations just as obesity and diabetes did, and the journals began reporting that as well.

Even when investigators compared similar aboriginal populations living in slightly different circumstances — as Frank Lowenstein, a medical officer for the World Health Organization, did with two tribes of Brazil Indians in the spring of 1958, one living on the grounds of a Franciscan mission and being fed by the missionaries, and one living isolated, deep in the rainforest — the population that was more acculturated had the higher blood pressure and the blood pressure that rose with age. When Lowenstein reviewed the medical literature of all such studies until then, his conclusion was: "All those groups which showed no increase of mean blood-pressures with age during adult life represent relatively small homogeneous populations living under primitive conditions in relative isolation, more or less undisturbed by their contacts with civilization . . . and they live almost entirely on the natural foods of their environment." Many factors could have explained it, Lowenstein suggested, because many "life habits" changed with Westernization. But if it could be explained, whatever the explanation turned out to be, this would likely explain both the hypertension and the rise of blood pressure with age that the rest of us experience.

By the 1980s, when 150 researchers from around the world published what was then the largest epidemiologic survey ever done on blood pressure, this Western disease phenomenon was still clearly visible. These researchers had measured blood pressure in fifty-two communities around the globe, of which four were still what Lowenstein would have called "relatively small homogeneous populations living under primitive conditions in relative isolation" — the Yanomamo and Xingu Indians of Brazil, and rural populations in Kenya and Papua New Guinea. Not only did these four have by far the lowest blood pressures measured, but their blood pressure remained low as they aged — which was not the case in *any* of the other populations in the study — and hypertension was virtually nonexistent.

The study, published in 1988, was known as INTERSALT because it had been designed to test the hypothesis that salt raises blood pressure; as a result, the investigators focused exclusively on blood pressure and salt. To the nutrition community, salt was not just the prime suspect for driving up blood pressure, but effectively the only one.* The same four isolated aboriginal

* In the 1960s, as the salt hypothesis took hold, researchers studying the rise of blood pressure with Westernization among nomadic tribes in Kenya and Uganda and South Pacific Islanders

populations that consumed relatively little salt also consumed relatively little sugar, but the investigators were interested in salt alone, as they had been since the 1960s.

The salt hypothesis has always been relatively simple and founded on basic physiology: Our bodies work to maintain a stable concentration of sodium (salt is sodium chloride) in our blood. When we consume a lot of salt, our bodies retain more water to dilute the sodium to the right concentration, and this manifests itself as elevations in blood pressure. Certainly in the short term, eating salt-rich snacks will make us thirsty, which is why bars and saloons typically offer such snacks for free, so they can sell us more of the liquids necessary to quench our thirst. Our kidneys are supposed to work by excreting the excess water and the salt in our urine, but the assumption is that they eventually fail to compensate, and chronically higher blood pressure is the result. Since the 1950s, this has been

---

first identified sugar and maybe white flour as the obvious culprits, because they were the conspicuous additions to the Westernized diets. However, the researchers switched their focus to salt when they realized that investigators in the United States were convinced that salt was the problem.

the standard thinking about the cause of hypertension, and the medical literature since then is also replete with dozens of randomized trials testing the hypothesis. ("As soon as we think we are right about something," the *New Yorker* writer Kathryn Schulz noted in her 2010 book *Being Wrong,* "we narrow our focus, attending only to details that support our belief, or ceasing to listen altogether.")

As with saturated fat and heart disease, though, this salt/hypertension hypothesis has resolutely resisted confirmation in clinical trials. For those not hopelessly wedded to the hypothesis, it has become increasingly difficult to believe that consuming too much salt is why we become hypertensive and why our blood pressure rises inexorably with age. Systematic reviews of the evidence from these trials invariably conclude that reducing our average salt intake by half, for instance, which is difficult to accomplish in the real world, will decrease blood pressure by 4 to 5 mm Hg mercury, on average, in those with hypertension, and perhaps 2 mm Hg in those without (known as normotensives). But even stage 1 hypertension, the less severe form of the condition, is defined by having a blood pressure elevated by at least 20 mm Hg over what's considered healthy. Stage 2 is defined as blood pressure elevated by at least 40 mm Hg over healthy levels. Hence, the fact that halving our salt consumption will result in a decrease of only 4 to 5 mm Hg suggests that the salt we eat is not the primary dietary driver of this disorder. This hasn't prevented public-health authorities from continuing to disseminate the message that salt is a "deadly white powder," as the Center for Science in the Public Interest hyperbolically phrased it in 1978. Avoiding the implications of these trials — that salt is not the cause of hypertension — has directed the research attention away from the possibility that something else in our diets or lifestyle is. If not salt, then what?

Not surprisingly, there's a long history of evidence implicating sugar — now in the laboratory and the clinic, as well as in the study of populations. As early as the 1860s, the German nutritionist Carl von Voit, a legendary figure in nutrition research, had suggested that something about eating carbohydrates made the human body retain water, which was not the case when fats are consumed. Francis Benedict, director of the Nutrition Laboratory at the Carnegie Institute of Washington, confirmed this observation in 1919 in one of the many seminal reports he and his Carnegie colleagues published.

By 1933, insulin was being implicated in this process, although the diabetes researchers at Columbia University who did so seemed unaware of the greater dietary context. Put simply, insulin seems to work as the opposite of a diuretic. Rather than promote the production of urine, which is what a diuretic does, it suppresses it, with the ultimate result being very similar to what is supposed to happen when we eat salt-rich foods. Insulin disturbs what is technically known as "electrolyte balance" or "electrolyte physiology" (sodium is an electrolyte) in such a way that the kidneys retain both sodium and water, rather than excrete them in the urine (just as insulin signals the kidneys to retain uric acid, and so plays a role in gout). By the 1950s, researchers were studying this phenomenon and publishing papers with titles like "Antidiuresis Associated with Administration of Insulin." Within another decade, the underlying biology of the phenomenon and insulin's effect on the kidneys, sodium retention, and thus hypertension had been elucidated. It was clear, in the words of the University of Texas endocrinologist Ralph DeFronzo, a pioneer with Gerald Reaven on the science of insulin resistance and metabolic syndrome, that "insulin, working through sodium, plays an important contributory role" in hypertension, particularly in individuals who happen to be obese and/or diabetic, and therefore insulin-resistant.

In the 1980s, Lewis Landsberg, a Harvard endocrinologist who would later become dean of the Northwestern University School of Medicine, discovered yet another mechanism by which insulin works to increase blood pressure and perhaps induce hypertension — in this case, by stimulating the central nervous system. Landsberg's revelation has since been integrated into established thinking as an explanation for why the obese are hypertensive: they're insulin-resistant, with chronically elevated levels of insulin, which in turn stimulates the nervous system, increasing heart rate, constricting blood vessels, and chronically elevating blood pressure. Since the obese seem to have increased sympathetic nervous activity, it makes perfect sense. Unfortunately, the medical community has continued to view this science as relevant only to the hypertension of the obese and diabetic; discussions on the dietary cause of hypertension have continued to focus almost obsessively on how much salt we should or should not be eating.

All these mechanisms by which insulin can elevate blood sugar and thus conceivably cause hypertension are directly relevant to the effect of sugar as well. If sugar causes insulin resistance and chronically elevates levels of insulin, then these are among the mechanisms through which it would be *expected* to cause hypertension. Richard Johnson's work on the fructose component of sugar and its effect on uric acid provides yet another, more direct means by which sugar would raise blood pressure. Johnson's research suggests that elevated levels of uric acid (at least in laboratory animals) leads to mild kidney damage and accelerates the process of kidney disease that's already established. The uric acid appears to cause the blood vessels in the kidneys to constrict and increases the blood pressure in the small capillaries (known as glomeruli) through which the kidneys filter waste products from the blood.

This, regrettably, links fructose and sugar not just to hypertension but to the kidney disease that is considered one of the "vascular complications" of diabetes, making it also a Western disease (albeit not mentioned in Burkitt and Trowell's provisional list). If

Johnson's work and its implications are correct, simply raising uric acid levels is enough to cause insulin resistance and thus, perhaps, type 2 diabetes and obesity, independent of these other effects on insulin and insulin resistance. And because the glucose in sugar appears to increase the rate at which we absorb and metabolize fructose, the two together — as in sucrose and high-fructose corn syrup — may indeed be the worst of all possible connections.

A final word about hypertension: When researchers study the effect of salt restriction on blood pressure in clinical trials, one possible explanation for the small overall effect these trials report is that some people may be particularly salt-sensitive, and others are not. Salt sensitivity is an elusive and controversial concept, but it implies that only some of us are sensitive to the salt content of the diet. For those of us who are, our blood pressure goes up and down in response to how much salt we're eating. Others can eat salt with impunity and their blood pressure remains relatively constant. That only some of us may be salt-sensitive is still considered by the public-health authorities reason enough to tell *everyone* to eat less salt. Their assumption is that those of us who are salt-sensitive will benefit and the rest will not be harmed. But salt sensitivity also seems to be associated with insulin resistance and metabolic syndrome. Salt-sensitive hypertension, for instance, can be caused in rats merely by damaging the capillaries of the kidney in the same way that high levels of uric acid do.

These observations and others have led researchers to suggest that salt sensitivity is caused by insulin resistance. If so, then telling people with or without salt-sensitive hypertension to eat less salt might ameliorate one of the symptoms of insulin resistance and metabolic syndrome — the hypertension. They would be better served by being told to avoid whatever was causing the insulin resistance and metabolic syndrome in the first place — i.e., sugar. That would take care of the root cause of the disorder, not just one of the symptoms.

Among the most provocative of the implications of the sugar/ insulin-resistance hypothesis is that cancer may well be caused or exacerbated by sugar. The supposition starts with two observations, the first of which is that cancer seems very much to be a disease of Western diets and lifestyles, just as Burkitt and Trowell suggested in their provisional list, and to increase in prevalence as populations become Westernized. The very concept of a disease of civilization begins with cancer. In 1844, Stanislas Tanchou, a French physician, a veteran of Napoleon's army and a knight of Napoleon's Legion of Honor, reported on his assessment of the death registries throughout Europe, concluding that cancer was more common in cities than in rural areas and that its incidence was increasing throughout the Continent. He acknowledged that cancer was an ancient disease, perhaps always present but, "like insanity," he famously said, it "seems to increase with the progress of civilization." Tanchou may have been the first of what would be a century of physicians, statisticians, and epidemiologists to poll physicians in distant and out-of-the-way locales, only to have them respond that diseases were rarely seen in their patient populations, or at least had been very rare occurrences, but were becoming more common with the passing of the years.

In 1902, the British government founded the Cancer Research Fund* to work with both the Royal College of Physicians and the Royal College of Surgeons in investigat-

* Later to be called the Imperial Cancer Research Fund and, today, Cancer Research UK.

ing "all matters connected with, or bearing on, the causes, preventions, and treatment of Cancer and Malignant Disease." The implicit message was that cancer appeared to be an increasingly common disease, and that action had to be taken to understand what was happening and why. A committee of investigators would now carefully examine the records of malignant disease in hospitals throughout the U.K., Europe, and Asia, and in missionary and colonial hospitals throughout the British Empire. A series of dispatches were circulated to the governors and commissioners of all the British colonies and protectorates worldwide, directing missionary and colonial physicians to report back on the prevalence of cancer in their patient populations and, if possible, ship specimens of any cancers that might be newly diagnosed and surgically removed ("placed in formalin immediately after removal from the body") back to London for careful microscopic investigation.

Within months, the letters and specimens began to arrive. Physicians responded from Newfoundland, the Caribbean, throughout Australia, New Zealand, and the South Pacific, from all the British protectorates in Africa, from the Mediterranean (Gibraltar and Malta), the Indian Ocean (Mauritius), and Asia. The replies reiterated a common theme: "There is a general unanimity of opinion in favor of the idea that cancer is a rare disease among the aboriginal tribes," a Dr. R. U. Moffat wrote about Kenya and Uganda, where he had worked first for the Imperial British East Africa Company and then the British government. Moffat had worked in East Africa for a decade, he reported, and yet had seen only "one undoubted case of cancer": a breast cancer in a Swahili woman living in Mombasa. (She refused an operation, he wrote, and her subsequent history was unknown.)

By 1908, when the fund's committee of cancer researchers and statisticians published its third report on its findings, a few relevant conclusions stood out. First, cancer incidence was definitely increasing across Europe, but it did so along with an "almost universal endeavor to improve the accuracy of statistics." Hence, it was impossible to determine whether or not cancer was, indeed, more frequent or whether physicians were merely paying it more attention and so more likely to diagnose and identify it when it did occur. Second, no population seemed to be exempt from cancer, but it was still undeniably rare in aboriginal or indigenous populations — in "the savage races," as the report put it. Although whether this was because the cancers weren't being diagnosed, or whether these people didn't live long enough to get cancer, or didn't go to these British doctors when they did, could also not be established. (Maybe they lacked what Joslin and Reginald Fitz had suggested about diabetics in the United States in 1898: the "wholesome tendency . . . to place themselves under careful medical supervision.")

The report concluded that it would "serve no useful purpose at present" to pursue the question further. But the question would not go away. In 1910 and again in 1915, researchers reported the results of surveys of Bureau of Indian Affairs physicians attending to Native American populations throughout the Midwestern and Western states. Both surveys concluded that cancer diagnoses and deaths among Native Americans served by these physicians were remarkably low, even though the Native Americans were apparently living at least as long as, if not longer than, the local whites. This relative absence of cancer, particularly breast cancer, was still the case more than half a century later, when Indian Health Service physicians began to survey medical records diligently among these Native

American populations.

When the American Cancer Society was founded in 1913 as the American Society for the Control of Cancer, it, too, carried out a systematic investigation with an expert committee led by Frederick Hoffman, formerly the chief statistician for Prudential Insurance. Hoffman published his seven-hundred-plus-page report *Mortality from Cancer Throughout the World* in 1915, concluding that far too many "qualified medical observers" were making this same observation — the relative absence of cancer in aboriginal and indigenous populations — and doing so in far too many locations around the globe to allow it to be explained away.

"There are no known reasons why cancer should not occasionally occur among any race or people, even though it be of the lowest degree of savagery or barbarism," wrote Hoffman. "Granting the practical difficulties of determining with accuracy the causes of death among non-civilized races, it is nevertheless a safe assumption that the large number of medical missionaries and other trained medical observers, living for years among native races throughout the world, would long ago have provided a more substantial basis of fact regarding the frequency of occurrence of malignant disease among the so-called 'uncivilized' races, if cancer were met with among them to anything like the degree common to practically all civilized countries."

Hoffman's report also concluded that cancer was that rare disease for which prevalence and mortality seemed to be steadily increasing — "one of the few diseases actually and persistently on the increase in practically all of the countries and large cities for which trustworthy data are obtainable." Hoffman and his colleagues estimated that cancer mortality in the United States had been increasing steadily by 2.5 percent per year. As with diabetes, this observation of increasing prevalence would be accompanied by a vigorous debate about whether or not those increases could be explained solely by the aging of the population, by new diagnostic techniques, by an increased tendency to attribute a death to cancer rather than old age or some other disease, or whether it was really the incidence and prevalence of cancer itself that was increasing.

Far more recent reports have concluded that it was, at least in part, the latter. "By the 1930s," as a 1997 report by the World Cancer Research Fund and American Institute of Cancer Research explained, "it was apparent that age-adjusted death rates from cancer were rising in the USA." This means that the likelihood of any particular sixty-year-old, for instance, dying from cancer was increasing, even if there were, indeed, more sixty-years-olds with each passing year. Some of this, of course, was due to the dramatic increase in lung cancers that in turn was a product of the epidemic of cigarette smoking that was aided and abetted by sugar. But this was true for cancers not related to smoking as well.

As for the evidence that cancer was a Western disease, this, too, continued to accumulate and remained a common observation through the 1930s. Among those who made it was Albert Schweitzer, who won the Nobel Peace Prize in 1952 for his missionary work. Schweitzer began working at a hospital in the equatorial lowlands of West Africa in 1913 and was, he later said, "astonished to encounter no cases of cancer" among the thousands of native patients he saw each year. However, as "the natives [took to] living more and more after the manner of the whites," he wrote, cancer in his patient population became ever more frequent.

After the Second World War, these observations are less common in the literature, but they don't vanish. In the 1950s, John Higginson, an American physician trained in England, surveyed cancer prevalence in native African populations and reported that it was still remarkably low compared with what was being reported in the United States and Europe. This led him to the conclusion that *most* human cancers are caused primarily by some aspect of diet and lifestyle. Because of this research and its implications, Higginson became, in 1965, the founding director of the World Health Organization's International Agency for Research on Cancer (IARC). In 1964, the WHO was suggesting that some proportion of human cancers, perhaps most, are "potentially preventable."

As late as 1952, malignant cancer among the Inuit was still deemed sufficiently rare that physicians working in northern Canada, as in Africa earlier in the century, would publish single-case reports in medical journals when they did diagnose a case. In 1984, Canadian physicians published an analysis of thirty years of cancer incidence among the Inuit in the western and central Arctic. Lung and cervical cancer had shown a "striking increase" over that time period, they reported, but there were still "conspicuous deficits" in breast-cancer rates. They could not find a single case of breast cancer in an Inuit patient before 1966; they could find only two cases between 1967 and 1980. Since then, breast-cancer prevalence has steadily increased among the Inuit, although it's still significantly lower than in other North American ethnic groups.

From the 1950s onward, popular thinking on the link between Western lifestyles and cancer focused on industrialization and carcinogens in the environment — something Higginson himself argued against in the 1980s, noting that "only a very small part of the total cancer burden" could be laid on industrial chemicals. When cancer epidemiologists did systematic reviews of the data, they continued to conclude, as Higginson had, that some significant percentage of cancers had to be lifestyle- or diet-induced. Breast cancer may be the best example. Though it has never been the scourge among Japanese women living in Japan that it is among women in America, it takes only two generations in the United States before Japanese-Americans experience the same breast-cancer rates as any other ethnic group. This implies that something about the American lifestyle or diet is a cause of breast cancer, although it doesn't tell us what that something is.*

In 1981, when the Oxford University researchers Richard Peto and Sir Richard Doll (knighted for his work linking cigarettes to lung cancer in the 1950s) published what was then the seminal article on cancer epidemiology, they estimated that perhaps three out of every four cases of cancer in the United States might be preventable with appropriate changes in diet and lifestyle. Diet, they argued, seemed to play the largest role. According to Peto and Doll's analysis, at least 10 percent of all cancers, and perhaps as much as 70 percent, were caused by something that we were eating.

The link between cancer and Westernization had taken on a new form by the early years of this century: the critical observation that obesity and diabetes both associate with an increased risk of cancer. The potential of such an association had been discussed in the medical literature as far back as the late nineteenth century — "the

* Not surprisingly, very similar patterns have been reported in other Western diseases as well — heart disease, for instance, as the epidemiologists Michael Marmot and Leonard Syme, then of the University of California, Berkeley, documented in 1976.

coincidence of diabetes and neoplasms [i.e., malignant tumors] . . . does not appear to be rare," as one 1889 article in the *British Medical Journal* phrased it — but it wasn't until the early years of this century that cancer researchers began to pay it serious attention.

In 2003, epidemiologists from the Centers for Disease Control, led by Eugenia Calle, published an analysis in *The New England Journal of Medicine* reporting that cancer mortality in the United States was clearly associated with obesity and overweight. The heaviest men and women, they reported, were 50 and 60 percent more likely, respectively, to die from cancer than the lean. This increased risk of death held true for a host of common cancers — esophageal, colorectal, liver, gallbladder, pancreatic, and kidney cancers, as well as, in women, cancers of the breast, uterus, cervix, and ovary. In 2004, the CDC followed up with an analysis linking cancer to diabetes, particularly pancreatic, colorectal, liver, bladder, and breast cancers. Cancer researchers trying to make sense of this association would later say that something about cancer seems to thrive on the metabolic environment of the obese and the diabetic.

One conspicuous clue as to what that

something might be was that the same association was seen with people who weren't obese and diabetic (or at least not yet) but suffered only from metabolic syndrome and thus were insulin-resistant. The higher their levels of circulating insulin, and that of a related hormone known as insulin-like growth factor, the greater the likelihood that they would get cancer. This link between cancer and insulin was evident with anti-diabetes drugs as well. In 2005, Scottish researchers reported that diabetic patients who took a drug called metformin, which works to reduce insulin resistance and therefore lower circulating levels of insulin, also had a significantly reduced risk of cancer compared with diabetics on other medications. That association has been confirmed multiple times, and has led researchers to test whether metformin acts as an anti-cancer drug, preventing or inhibiting cancer's recurrence in randomized controlled trials. These observations also served to focus the attention of cancer researchers further on the possibility that insulin and insulin-like growth factor are cancer promoters, and thus that abnormally elevated levels of insulin — caused by insulin resistance, for instance — would increase our cancer risk.

This was another area of research that had emerged in the 1960s, with laboratory work by some of the leading cancer researchers — including Howard Temin, who would later win the Nobel Prize — demonstrating that cancer cells require insulin to propagate; at least they do so outside the human body, growing as cell cultures in the laboratory. This would turn out to be the case for breast-cancer cells, even though the normal breast cells from which these malignant cells emerged lacked insulin receptors and lacked the necessary machinery within the cells to respond to insulin signaling. Nevertheless, as the University of Toronto cancer researcher Vuk Stambolic would later describe it, these breast-cancer cells seemed to be "addicted to" insulin, and when weaned off it in the laboratory they responded by dying. This kind of phenomenon was seen also in cancers of adrenal and liver cells. As one 1976 report put it, insulin "intensely stimulated cell proliferation in certain tumors"; another, by researchers at the National Cancer Institute, described one particular line of breast-cancer cells as "exquisitely sensitive to insulin." By then, researchers had established that malignant breast tumors had receptors to insulin, which were absent in healthy breast tissue, and that the more they had, the more insulin-sensitive they were.

Insulin-like growth factor (IGF) was discovered only in the 1950s; as its name implies, it has a structure very similar to that of insulin and its effect on cells can mimic that of insulin. But IGF is secreted in response to growth hormone, rather than carbohydrate or protein consumption, as insulin is. It's also secreted in response to insulin itself. Tumor cells appear to have two to three times the amount of IGF receptors as normal cells, and researchers believe that functioning IGF receptors are necessary for the growth of cancer cells. The consensus among researchers studying the role of insulin and IGF in cancer is that these hormones supply both the fuel necessary for tumors to divide and multiply, and provide the signals necessary to the tumors to keep doing so. The more insulin and IGF in the circulation, the more cancer cells are driven to multiply and tumors to grow.

The science on the link between insulin and IGF and cancer now has been well worked out. A consensus has been forming, led by some of the most respected cancer researchers — in particular Lewis Cantley, who runs the cancer research program at Weill Cornell Medical College, and Craig

Thompson, president of the Memorial Sloan Kettering Cancer Center, both in New York City. These researchers believe that cancer is as much a metabolic disease as a "proliferative" disease, and that for cancer cells to procreate, they have to rewire their metabolic programs — how they fuel themselves — to drive their unfettered growth. Further evidence to support this view is that the major genetic mutations that have been discovered over the years as seemingly responsible for a host of different cancers seem to play critical roles, not just in the proliferation of cells but in regulating the metabolism of the cells.

From this perspective of cancer as a metabolic disease, insulin and IGF promote the cancer process through a series of steps. First, insulin resistance and elevated levels of insulin trigger an increased uptake of blood sugar (glucose) as fuel for precancerous cells. These cells then begin producing energy through a mechanism known as aerobic glycolysis that is similar to what bacteria do in oxygen-poor environments. (This phenomenon is known as the Warburg effect and was discovered in the 1920s by the German biochemist and later Nobel Laureate Otto Warburg, although its importance in the cancer process was not embraced until recently.) Once cancer cells make this conversion, they burn enormous amounts of glucose as fuel, providing them, apparently, with the necessary raw materials to proliferate.

By metabolizing glucose at such a rapid rate, as Thompson suggests, these cancer cells generate relatively enormous amounts of compounds known technically as "reactive oxygen species" and less technically as "free radicals," and these, in turn, have the ability to mutate the DNA in the cell nucleus. The more glucose a cell metabolizes and the faster it does so, the more free radicals are generated to damage DNA, explains Thompson. And the more DNA damage, the more mutations are generated, and the more likely it is that one of those mutations will bestow on the cells the ability to proliferate without being held in check by the cellular processes that work to prevent this pathological process in healthy cells. The result is a feed-forward acceleration of tumor growth. While this is happening, the insulin and IGF in the circulation both work to signal the cell to keep proliferating, and to inhibit the mechanism (technically known as apoptosis, or cell suicide) that would otherwise kick in to shut it down.

These researchers can imagine two ways

in which insulin and IGF are involved in the initiation of the cancer process based on the understanding that has emerged in the last decade.

One is for mutations to occur in the DNA of our cells — by bad luck, in effect — which work to increase the strength of the signal that insulin and IGF send to cells and thus make the cell take up more glucose and start on the road to cancer. Because this doesn't actually require insulin resistance and high levels of insulin in the bloodstream, these cancers, to borrow a term from the diabetes literature, would be non-insulin-dependent. They would grow and propagate even when insulin levels are low and the host (i.e., the person in the process of getting cancer) is insulin-sensitive.

But the other way to initiate the cancer process, according to these researchers, is to increase the levels of insulin and blood sugar in the circulation itself. Insulin resistance would do that. Thus whatever is causing insulin resistance would be promoting the transformation of healthy cells into malignant, metastatic cells by increasing insulin secretion and elevating blood sugar and telling the cells to take up increasingly more glucose for fuel.

This leads those like Cantley and Thompson directly back to sugar. As Cantley has said, sugar "scares" him, for precisely this reason. If the sugars we consume — sucrose and HFCS specifically — cause insulin resistance, then they are prime suspects for causing cancer as well, or at the very least promoting its growth. Even if the details of the mechanism should turn out to be wrong, the association between obesity, diabetes, and cancer, and the specific association between insulin, IGF, and cancer, suggests that whatever is causing insulin resistance is increasing the likelihood that we will get cancer. If it's sugar that causes insulin resistance, it's hard to avoid the conclusion that sugar causes cancer, radical as this may seem, and even though this suggestion is rarely if ever voiced publicly.

By now, the message should be clear: if insulin is involved in a disease process, then insulin resistance — i.e., metabolic syndrome — is likely to make it worse, and perhaps even initiate the disease process to begin with. This directly implicates sugar as a potential cause, a dietary trigger of the disease.

Dementia has a long history, and we're unlikely ever to answer the question of whether it is more common now than it once was. The risk of getting Alzheimer's disease roughly doubles every five years past the age of sixty — or at least it does in modern Western societies — and so, the longer a population lives, the greater the burden or prevalence of Alzheimer's. Since we happen to be living considerably longer than our ancestors, our risk is increasing.

The pathological signature of Alzheimer's disease was only officially recognized in the early years of the twentieth century — the association of a rapidly deteriorating dementia with the distinctive accumulation in the brain of what are called amyloid plaques and neurofibrillary tangles. As historians of medicine have noted, however, the plaques and tangles had been previously identified. But Alois Alzheimer happened to have personal experience with the relatively young demented patient in whose postmortem brain he observed these phenomena in 1906. Alzheimer's name was then attached eponymously to the disease, not necessarily because it was a new or rare disease (although it might have been), but because the head of the institute at which Alzheimer was doing his research apparently wanted to claim that it was. Although several studies have compared the prevalence of Alzheimer's disease in various populations and suggested that it might be a product of Western diets and lifestyles, this evidence is not nearly as clear is it is with diabetes or even cancer.

Alzheimer's, like cancer, is associated with type 2 diabetes, an observation that began to emerge from studies in the mid-1990s of eight hundred elderly residents of Hisayama, Japan; of seven thousand senior citizens in Rotterdam, the Netherlands; and of fifteen hundred type 2 diabetics in Rochester, Minnesota. These observations have been confirmed repeatedly since. They suggest that type 2 diabetics have from one and a half to two times the risk of Alzheimer's dementia of nondiabetics, suggesting in turn, as the Rotterdam investigators did in 1999, that "direct or indirect effects of insulin could contribute to the risk of dementia." Waist circumference is also associated with Alzheimer's risk — the thicker your waist, the greater your risk — as is Body Mass Index itself, although only in midlife, not afterward. Getting fatter (as many of us do) in our thirties and forties is associated with an increased risk. Several studies have shown that higher insulin levels — hyperinsulinemia — are associated with increased risk. Hypertension is also associated with increased risk of Alzheimer's.

Over the years, researchers have suggested numerous possibilities to explain these associations, covering the entire range of metabolic and hormonal disorders that accompany type 2 diabetes. Perhaps the high blood sugar (glycemia) is responsible for the increased risk of Alzheimer's disease; the higher the blood sugar, the greater the oxidative stress in the brain, and the greater the production of what are called advanced glycation end products, AGEs. These AGEs are associated with the accumulation of plaques and tangles and may have a causative role. Maybe it's the hypertension itself. Maybe the inflammation that seems to accompany obesity is responsible, and thus the "inflammatory" molecules that overstuffed fat cells will secrete.

Researchers have now unraveled a host of mechanisms by which insulin plays a role in the brain that could go awry with insulin resistance in ways that might either cause or exacerbate the Alzheimer's process. This thinking has led some researchers to think of Alzheimer's as type 3 diabetes, because of the possibility that it is intimately related to insulin signaling and insulin resistance. In a 2014 review article, C. Ronald Kahn, a former director of the Joslin Diabetes

Center, and two colleagues from Harvard Medical School enumerated the multiple ways identified so far in which insulin signaling in the brain "is vital in the fine-tuning of brain activity." They then discussed the many mechanisms by which dysregulation of this insulin signaling can lead to both cognitive and mood disorders and to Alzheimer's disease. These include direct impairment of the function of neurons and what is called "synaptogenesis" (the formation of synapses — i.e., connections — between neurons, which goes on throughout our lives and is critical to healthy brain functioning), as well as mechanisms that work more directly to increase the rate at which plaques and tangles accumulate in the brain, or decrease the rate at which the brain can clear away these pathological phenomena. All of this is still speculative, but there's another major factor involved in the association of type 2 diabetes and Alzheimer's that is considerably less so.

Alzheimer's disease is by no means the only possible cause of dementia, nor is it the only one strongly associated with age and with type 2 diabetes. Both type 2 diabetes and hypertension clearly increase our risk of cerebrovascular disease and stroke — a blockage in the blood vessels in the brain (hence a "cerebrovascular accident") — which cuts off the blood supply to a portion of the brain. The result is the death of brain tissue (an "infarct" or a "microinfarct") and, depending on the location and extent of the damage, dementia. This is what is known technically as vascular dementia. When confronted with a patient suffering from dementia, physicians may likely diagnose vascular dementia, based on the observation that the dementia itself followed closely on the heels of a stroke and was not the kind of gradual decline seen typically in Alzheimer's. But this is an oversimplification of the process.

Among the seminal findings in dementia research over the past twenty years is that we all tend to accumulate plaques and tangles in the brain as we age, as well as some degree of vascular damage, whether we manifest dementia or not. The plaques and tangles remain the classic pathological signatures of Alzheimer's disease, but the more vascular damage that accumulates — the infarcts and microinfarcts — the lower the threshold for dementia to appear. This was first observed in a seminal study of nuns in the Sisters of Notre Dame congregation that was published in 1997 by University of Kentucky researchers, and it has been confirmed in studies since then. These studies conclude that for any given amount and distribution of plaques and tangles in the brain, the more vascular damage that is also present, the more likely we are to appear demented and to be diagnosed on autopsy as having had Alzheimer's disease, simply because the physician making the diagnosis will be more aware of the dementia. Depending on a host of factors, genetics being one of them, this will happen to some of us faster than others. When we cross some threshold of damage, dementia begins to manifest itself. If we're diabetic and hypertensive, which also means we're insulin-resistant, we're going to have more vascular damage and so reach that threshold of damage sooner.

This will happen whether or not insulin or insulin resistance is involved directly in the Alzheimer's disease process. And, once again, it implies that if sugar causes the insulin resistance, and thus the type 2 diabetes and the hypertension, then sugar also increases the likelihood that dementia is in our future.

Here's another way to think about the idea that a cluster of chronic Western diseases associate with insulin resistance, metabolic syndrome, obesity, and diabetes and hence sugar consumption: Diabetes, though a discrete diagnosis by our doctors, is not a discrete phenomenon in which bad things suddenly start happening that didn't happen before. It's part of a continuum from health to disease that is defined in large part by the worsening of the metabolic abnormalities — the homeostatic disruption in regulatory systems — that we've been discussing and that are associated with insulin resistance, if not caused by it, and so part and parcel of metabolic syndrome.

As we become ever more insulin-resistant and glucose-intolerant, as our blood sugar gets higher along with our insulin levels, as our blood pressure elevates and we get ever fatter, we are more likely to be diagnosed as diabetic and manifest the diseases and conditions that associate with diabetes. These include not just heart disease, gout, cancer, Alzheimer's, and the cluster of Western diseases that Burkitt and Trowell included in their provisional list, but all the conditions typically perceived as complications of diabetes: blood-vessel (vascular) complications that lead to strokes, dementia, and kidney disease; retinopathy (blindness) and cataracts; neuropathies (nerve disorders); plaque deposits in the arteries of the heart (leading to heart attacks) or the legs and feet (leading to amputations); accumulation of advanced glycation end products, AGEs, in the collagen of our skin that can make diabetics look prematurely old, and that in joints, arteries, and the heart and lungs can cause the loss of elasticity as we age. It's this premature aging of the skin, arteries, and joints that has led some diabetes researchers to think of the disease as a form of accelerated aging. But increasing our risk of contracting all these other chronic conditions means we're also likely to get these ailments at ever-younger ages and thus, effectively, age faster.

A host of other pathological phenomena also associate with metabolic syndrome and insulin resistance. Researchers have typically studied these from the perspective that they are somehow caused by getting fatter, by eating too much or exercising too little, or maybe even by eating too much fat. These phenomena work to trigger hyperinsulinemia and insulin resistance. Fat, as we've discussed, accumulates in our livers and muscle cells, a process these researchers refer to as lipotoxicity. Stress hormones (cortisol, for instance) increase in the circulation; inflammation increases, as signified by the increase in our circulation of inflammatory molecules (secreted by fat cells). More reactive oxygen species (free radicals) are generated, and so oxidative stress increases. The mitochondria in our cells become dysfunctional. For virtually all of these, as the researchers will acknowledge if they're being suitably skeptical, "the direction of the relationship is still unclear: it may be a cause or consequence of insulin resistance." All of this is happening coincident with the development of insulin resistance and metabolic syndrome, and all of it gets worse as we become fatter and more diabetic. All of this has pathological effects throughout our bodies. All of this is triggered by something in our diet and lifestyle, which is what we ultimately have to explain.

Another issue that has recently added still another layer of complication to the science is the role played in obesity and diabetes by the bacteria in our guts, known as the gut microbiota or microbiome. New technologies will lead inevitably to new areas of research, new observations, and new discoveries. The ability to sequence the genomes of these bacterial species has opened up a new frontier of research, just as the ability to measure blood pressure, cholesterol, or insulin sensitivity did for earlier generations of researchers. The microbiome research, because it's brand-new, is at a very preliminary stage.

Still, as the new new thing (to borrow a phrase from the journalist Michael Lewis) in obesity and diabetes research, gut bacteria get an inordinate amount of attention, particularly from the media, though we may not know for decades what to make of the observations that ensue — what is signal and what is noise. Most of the work so far has been done in laboratory mice and rats, and the relevance to human life (or even to other laboratory animals) is unclear. The observations that come from human studies and the very few human experiments are still impossible to interpret reliably. Certain alterations in this gut microbiome associate with obesity, metabolic syndrome, and diabetes, but, as the researchers will acknowledge, "it remains to be determined whether these are the results of altered glucose metabolism and insulin resistance or contribute to their development."

Since the 1950s, if not earlier, researchers have known that the foods we eat and the form in which they come — indigestible fiber, refined grains and sugar, and all the rest — will influence which species of gut bacteria thrive and which don't. That in turn will affect the digestibility of the fat, protein, and carbohydrates in the rest of our food and the effect on blood levels of cholesterol and triglycerides, if nothing else.

Ultimately, what we have to keep in mind as we read the latest articles on recent developments in the science is the critical observations that so desperately have to be explained: If specific changes in the bacterial species that populate our digestive tract associate with obesity and diabetes, this suggests that these changes are yet another effect of the same underlying cause. And the most likely suspect driving any related pathological changes in these bacterial populations would once again be the radical increases in sugar consumption that come with Western lifestyles. "It would be an extraordinary coincidence," as Peter Cleave wrote and we've already quoted, "if these refined carbohydrates, which are known to wreak such havoc on the teeth, did not also have profound repercussions on other parts of the alimentary canal during their passage along it, and on other parts of the body after absorption from the canal."

Nutrition researchers and public-health authorities have typically been of two minds about the hypothesis that a single nutrient might be to blame for this spectrum of chronic disease states that associates with insulin resistance, metabolic syndrome, obesity, and type 2 diabetes, or that a single phenomenon might be responsible.

On the one hand, as we've said, they've been willing to blame the victims, at least those who are overweight or obese, for eating too much and exercising too little, and the food industry for making too much food available and for manipulating the taste with sugar, salt, and fat to the point that we just can't eat in the necessary moderation. They've also entertained the possibility that dietary fat and particularly saturated fat plays a uniquely causal role. But their tests of this dietary fat hypothesis have mostly failed to support it.

Since the 1970s, though, they've considered it quackery to suggest that sugar is responsible. Since then, well over half a million articles have been published in the peer-reviewed medical literature on the subjects of obesity and/or diabetes, while the prevalence of those diseases in our society has inexorably climbed. The implication is that if this were a simple problem we surely would have solved it by now, so it must be multifactorial and complex — two words that are invoked so consistently to explain the genesis of these diseases that we have to question whether the terms imply an explanation or a simple lack of understanding of the problem.

The way we fund science in nutrition and chronic disease research is also partly responsible for this thinking. The confluence of diet and chronic disease is not a scientific discipline in which all or many of the researchers band together to answer a few critically important questions, although I would argue that it should be. The National Institutes of Health and other research agencies fund thousands or tens of thousands of researchers to answer thousands or tens of thousands of small questions, and the hope is that out of these pieces a coherent picture will emerge. Instead, what we have is a cacophony and the assumption that if so many researchers are studying so many different pieces of the puzzle, it must be a very complex problem.

More recently, journalistic authorities on the subject of food and health have also expressed their displeasure at "one nutrient" explanations for our ills. They perceive such explanations as overly simplistic, if not a kind of idealistic wishful thinking. This leads in turn to the notion that the industrialization of the food industry and the processing of most modern foods yield so many potentially deleterious changes that making sense of them all is beyond the realm of science to establish, and therefore we should, more or less, stop trying. As the University of California, Berkeley, authority Michael Pollan has so memorably put it, we should "eat food. Not too much. Mostly plants." If we do this, we will get as close as we reasonably can to a healthy diet.

But science is about explaining what we observe in nature and doing so with the simplest possible explanation — as Newton suggested, with the simplest explanation that is both true and sufficient. The process of science is then about the conflict between the desire to believe a simple explanation — particularly our simple explanation — and the skepticism required to establish reliably whether it does or does not explain what we observe.

Here we're back to those few observations that are indisputable and that we have to explain. In the second half of the nineteenth century in Western populations, and far more recently in others, obesity and type 2 diabetes emerged, eventually to become the dominant diseases of modern times. Insulin resistance characterizes both these disorders. And those who are insulin-resistant, who suffer from obesity and type 2 diabetes, are at higher risk of a host of other chronic diseases — the Western diseases, as Burkitt and Trowell described them — and these diseases, too, are associated with insulin resistance.

How do we explain these observations? What has changed that could cause the emergence of these diseases worldwide and the insulin resistance that is associated with so many of them? What changes in our diets and our lifestyles can explain these changes in disease patterns? Is a simple hypothesis sufficient to do it? Is it that we're all simply eating too much and exercising too little, which is the one simple answer that the nutritional establishment will embrace in the face of so much evidence to the contrary? Another simple answer, and a more likely one, is sugar.

Epilogue: How Little Is Still Too Much?

It's impossible to say. In 1986, when the FDA concluded that most experts considered sugar safe (at least at the annual level of forty-two pounds per capita that the FDA administrators decided we were then consuming), and when the relevant research communities settled on caloric imbalance as the cause of obesity and saturated fat as the dietary cause of heart disease, the clinical trials necessary to begin to answer such a question were never pursued.

The traditional response to the how-little-is-too-much question is that we should eat sugar in moderation — not eat too much of it. But this is a tautology. We only know we're consuming too much when we're getting fatter or manifesting other symptoms of insulin resistance and metabolic syndrome. At that point, the assumption is that we can dial it back a little and be fine — drink one or two sugary beverages a day instead of three, or, if we're parenting, allow our children ice cream on weekends only, say, rather than as a daily treat. But if it takes years or decades, or even generations, for us to get to the point where we manifest symptoms of metabolic syndrome, it's quite possible that even these apparently moderate amounts of sugar will turn out to be too much to reverse the situation and return us to health. And if the symptom or complication of metabolic syndrome and insulin resistance that manifests first is something other than getting fatter — cancer, for instance — we're truly out of luck.

The authorities (or self-appointed authorities) who argue for moderation in our eating habits tend to be those who are relatively lean and healthy; they define moderation as what works for them. This assumes that the same approach and amount will have the same beneficial effect on all of us (and that it will continue to work for them as well). If it doesn't, of course, if we fail to remain lean and healthy or our children fail to do so, the assumption that, naturally again, follows from this perspective is that we've failed in our assessment of moderation — we ate too much sugar or our children did.

To understand this tautological logic better, imagine a situation in which cigarette smokers who don't get lung cancer (or heart disease or emphysema) assume de facto that those smokers who do are those who smoke "too much." They'd certainly be right, but it still wouldn't tell us what constitutes a healthy level of smoking, or whether such a thing as smoking in moderation even exists. How many cigarettes could be smoked without doing at least some harm to our health, and could thus constitute smoking in moderation? If we say none, we may, indeed, be right, but now we've redefined how we're willing to work with the concept of moderation. The same logic may also apply to sugar. If it takes twenty years of either smoking cigarettes or consuming sugar for the consequences to appear, how can we know whether we've smoked or consumed too much before it's too late? Isn't it more reasonable to decide early in life (or early in parenting) that not too much is as little as possible?

Recall the thinking of Priscilla White, who went to work in 1924 with Elliott Joslin at his diabetes clinic in Boston and oversaw the treatment of the clinic's pediatric cases. "No child can grow up without a scoop of ice cream once a week," White had said, although the translation of this belief into clinical practice would require that the children who got their weekly scoop also had to inject more insulin over the course of their lives than children whose parents and doctors might have taken a stricter approach. Had White known (as she couldn't at the time) that eating a weekly scoop of ice cream *and* taking more insulin in response would make children suffer greater complications from their diabetes and die earlier than those who abstained from the ice cream, would that have influenced her thinking? I'd bet that it would have; I'd also bet that she would have wanted to know the increase in disease burden and decrease in longevity per scoop of ice cream consumed, if such a thing were possible — as would the parents — before deciding whether a scoop a week was "too much" for these children. And if these children never ate ice cream, would they miss it any more than would a child who never takes up the habit of smoking miss the opportunity as an adult to indulge occasionally in a cigarette?

Any discussion of how little sugar is too much also has to account for the possibility that sugar is a drug and perhaps addictive. Even if "people just act like it is," as Charles C. Mann has written, this suggests the possibility that having the opportunity to consume at least some sugar (or ice cream) is only meaningful in a world in which substantial sugar consumption is the norm and virtually unavoidable and everyone does it. Trying to consume sugar in moderation, however it's defined, in such a world is likely to be no more successful for some of us than trying to smoke cigarettes in moderation — just a few cigarettes a day, rather than a pack. Whether or not we can avoid any meaningful chronic effects by doing so, we may not be capable of managing our habits, or managing our habits might become the dominant theme in our lives (just as rationing sweets for our children can seem to be a dominant theme in parenting). Some of us certainly find it easier to consume no sugar than to consume a little — no dessert at all, rather than a spoonful or two before pushing the plate to the side. If sugar consumption may be a slippery slope, then advocating moderation is not a meaningful concept.

We can also try to define "too much" from a population perspective — perhaps too broadly, too myopically. George Campbell's estimate from the 1960s of seventy pounds of sugar per capita prior to the appearance of a diabetes epidemic may have been reasonable, and the assumption of the 1986

FDA report that forty-two pounds per capita is a safe amount may also have been, but the appearance of a diabetes epidemic and of diabetes itself are two different things. If the fuse of the diabetes epidemic is lit a generation or more before the epidemic explodes, if the predisposition to become insulin-resistant, obese, and diabetic is passed down and amplified from mother to child in the womb, then it becomes far more difficult to establish at what level of sugar consumption a population, let alone an individual, remains healthy, or becomes healthy again if they're not. What appears to be a population threshold of seventy pounds per capita yearly might actually be a threshold of thirty pounds a generation or two or three earlier. Once we've crossed the threshold and are on our way to becoming an obese and diabetic population, it's likely that we have become different physiologically, that the children in a population that has been consuming a significant amount of sugar for generations have been programmed differently to respond to a sugar-rich environment from those who were born earlier. There may be no going back, or not without drastic changes in our diet. The existing research provides no way to know.

In my own mind, I keep returning to a few observations — unscientific as they may be — that make me question the validity of any definition of moderation in the context of sugar consumption. One was the suggestion by Hindu physicians more than two thousand years ago that sugar consumption could promote both nutrition *and* corpulence and, as Frederick Allen noted, that diabetes might be brought on by eating sugar, partly because of the sweet smell of the urine and partly because diabetes then seemed to be a disease exclusively of the affluent, who alone could afford to indulge in sugar and flour. ("This definite incrimination of the principal carbohydrate foods," as Allen had written, "is, therefore, free from preconceived chemical ideas, and is based, if not on pure accident, on pure clinical observation.")

Then there was Thomas Willis in the 1670s, the first physician in Europe to note the sweet taste and smell of diabetic urine, despite a long tradition among European physicians at the time of tasting urine as a diagnostic technique. Why hadn't physicians noticed until then, primitive as the art of diagnosis might have been? Willis's identification of diabetes and the sweetness of the urine happens to coincide both with the first flow of sugar into England from its Caribbean colonies, and with the first use of sugar to sweeten tea, which was now being imported into England from China.

Other observations that resonate with me when I wrestle with the concept of moderation include one of Frederick Slare's comments in 1715 in his "Vindication of Sugars Against the Charges of Dr. Willis." At a time when sugar was just beginning to make its transition in England from Sidney Mintz's "luxury of kings into the kingly luxury," Slare noted that women who cared about their figures but were "inclining to be too fat" might want to avoid sugar, because it "may dispose them to be fatter than they desire to be." In a similar vein, the French lawyer-turned-gastronome Jean Anthelme Brillat-Savarin suggested in 1825 in *The Physiology of Taste,* perhaps the most famous book ever written about food, that obesity was caused by the consumption of starches and bread ("fecula" or "farinaceous foods," he called them) and that this fattening process occurs "more quickly and surely" when such foods are consumed with sugar. In the 1860s, the Portuguese physician Abel Jordão commented that sugar was likely to be a fattening agent, in turn prompting Charles Brigham at Harvard to observe that young women of his era, worried about the "skeleton-like appearance which their shoulders and arms present when exposed," had taken to consuming sugar water to put on some fat and appear more womanly.

In all these cases, even the affluent would likely have been consuming less sugar than Campbell's seventy-pound estimate or the FDA's forty-two. When Slare made his observation in 1715, the English were consuming, on average, perhaps five pounds of sugar a year.

Combine these observations with the research implicating high blood sugar and insulin resistance in the intrauterine environment — the influence of metabolic programming or imprinting on the generation to come — and it suggests that our consumption of sugar over the centuries may have changed the species. Transform an environment so dramatically — as sugar has transformed what we eat and drink in ours — and the species in that environment will be transformed as well. It suggests that the response of individuals today to any amount of sugar is vastly different from what it would have been centuries ago. Perhaps we can tolerate less, perhaps more; we can only speculate. Nor can we say how sugar consumption in a population over generations changes the pattern of chronic diseases that appear and work to shorten lives, and how that differs, as Denis Burkitt would have noted, in different populations with different genetics.

Imagine, for instance, a thought experiment: A population of individuals who have never consumed refined sugar in any quantity, other than what they eat naturally in fruits and vegetables. This population is split in two and then followed for generations. One population has access to refined sugar and high-fructose corn syrup and consumes them in ever-increasing quantities, and the other continues its relatively sugar-free existence. Both populations have access to the same advances in medical care and public health as the generations roll by. Do they both end up with the same spectrum of chronic diseases — similar levels of heart disease, diabetes, cancer, and dementia? And if the sugar-eating population, as I'm suggesting, has the far greater burden of chronic disease, and it is then taken off sugar, how many generations would have to go by before the two populations were again equivalent? Would they ever be?

That experiment can exist only in our imagination — in real life, all populations were put on the sugar-rich diet. Hence, we don't know what "normal" or "healthy" would have looked like in a sugar-free or even low-sugar world. We don't know what our species would have become. Would we get fat as we get older? Would our LDL cholesterol and triglycerides and blood pressure increase with age? Would we become ever more glucose-intolerant and resistant to the action of insulin? How long would we typically live? What diseases would ultimately kill us? These questions cannot be answered.

Imagining such an experiment also helps us understand why future research might never be able to resolve these questions definitively. This speaks to the point I raised earlier, acknowledging that the evidence against sugar is not definitive, compelling though I may personally find it to be. Let's say we randomly assigned individuals in our population to eat a modern diet with or without sugar in it. Since virtually all processed foods have sugar added or, like most breads, are made with sugar, the population that is asked to avoid sugar would simultaneously be avoiding virtually all processed foods as well. They would dramatically reduce their consumption of what Michael Pollan has memorably called "foodlike substances," and if they were healthier, there would now be a host of possible reasons why. Maybe they ate fewer refined grains of any type, less gluten, fewer trans fats, preservatives, or artificial flavorings? We would have no practical way to know for sure.

We could try to reformulate all these foods so that they are made without sugar, but then they won't taste the same — unless, of course, we replace the sugar with artificial sweeteners. Our population randomized to consume as little sugar as possible is likely to lose weight, but we won't know if it happened because they ate less sugar, or fewer calories of all sorts. Indeed, virtually all diet advice suffers from this same complication: whether you're trying to avoid gluten, trans fats, saturated fats, or refined carbohydrates of all types, or just trying to cut calories — eat less and eat healthy — an end result of this advice is that you're often avoiding processed foods containing sugar and a host of other ingredients. If we benefit, we cannot say exactly why. It is too complicated.*

* The diet that many public-health authorities believe is the healthiest is known as DASH — Dietary Approaches to Stop Hypertension. The authors of the first study on DASH described it as

Diet advice that recommends we eat whole foods and avoid processed foods (foodlike substances) removes virtually all refined sugars by definition; diet advice to avoid sugar means, by definition, that we avoid virtually all processed foods.

Artificial sweeteners (noncaloric sweeteners, as the USDA calls them) as a replacement for sugar muddy these waters even more. Much of the anxiety about these sweeteners was generated in the 1960s and 1970s by the research, partly funded by the sugar industry, as we've seen, that led to the banning of cyclamates as a possible carcinogen, and the suggestion that saccharin could cause cancer (at least in rats, at extraordinarily high doses). Though this particular anxiety has tapered off with time, it has been replaced by the suggestion that maybe these artificial sweeteners can cause meta-

---

"rich in fruits, vegetables, and low-fat dairy foods and with reduced saturated fat and total fat." A primary goal of this dietary prescription is to provide significant potassium, magnesium, and calcium, with the assumption that this in turn will lower blood pressure. But it also prohibits sugar, sweets, and sugary beverages other than fruit juices. Its benefits may come as much from that restriction as any other.

bolic syndrome, and thus obesity and diabetes.

This conjecture comes primarily from epidemiological studies that show an association between the use of artificial sweeteners and obesity and diabetes. But whether this means artificial sweeteners *cause* obesity and diabetes is, again, impossible to say. It is likely that people who are predisposed to gain weight and become diabetic are also the people who use artificial sweeteners instead of sugar. The latest review articles on the subject of possible dangers from artificial sweeteners suggest that the evidence is, indeed, far short of definitive. Though the possibility can't be ruled out that consuming artificial sweeteners will lead to increases in morbidity and mortality, it seems unlikely.

As Philip Handler, head of the National Academies of Sciences, suggested in 1975, or as President Teddy Roosevelt did in 1907, what we want to know is whether using artificial sweeteners over a lifetime — or even a few years or decades — is better or worse for us than however much sugar we would have consumed instead. It's hard for me to imagine that sugar would have been the healthier choice. But the research can say no more *definitively* about this question than it can about the long-term effects of consuming sugar. Laboratory research has identified mechanisms by which artificial sweeteners *might* trigger physiological responses in the body similar to those triggered by sugar. We have sweet-taste receptors in our guts and digestive tracts, as well as in our mouths, for instance, and so the same molecules that trigger these and fool the brain into thinking we're consuming sugar might fool the body as well. If it does, though, there's little evidence that it results in deleterious effects on food intake, metabolic syndrome, and body weight of the kind observed with sugar itself. If the goal is to get off sugar, then replacing it with artificial sweeteners is one way to do it. Whether consuming artificial sweeteners for years or decades brings on its own noxious effects, or prevents us from benefiting fully from a sugar-free diet, is something that the existing research cannot say.

The research community can certainly do a much better job than it has in the past of testing all these questions. But we may have a very long wait before the public-health authorities fund such studies and give us the definitive answers we seek. What do we do until then?

Ultimately and obviously, the question of

how much is too much becomes a personal decision, just as we all decide as adults what level of alcohol, caffeine, or cigarettes we'll ingest. I've argued here that enough evidence exists for us to consider sugar very likely to be a toxic substance, and to make an informed decision about how best to balance the likely risks with the benefits. To know what those benefits are, though, it helps to see how life feels without sugar. Former cigarette smokers (of which I am one) will tell you that it was impossible for them to grasp intellectually or emotionally what life would be like without cigarettes until they quit; that through weeks or months or even years, it was a constant struggle. Then, one day, they reached a point at which they couldn't imagine smoking a cigarette and couldn't imagine why they had ever smoked, let alone found it desirable.

A similar experience is likely to be true of sugar — but until we try to live without it, until we try to sustain that effort for more than days, or just a few weeks, we'll never know.

ACKNOWLEDGMENTS

The Case Against Sugar is my third book on nutrition and chronic disease. It is as much a product as the earlier two of my reporting on the subject from the late 1990s onward. I remain grateful and indebted to the many hundreds of researchers and public-health authorities who graciously gave of their time to be interviewed, and to the editors, readers, and research assistants who helped shape those earlier projects and make them possible.

This book had its genesis on January 23, 2008, when I received an e-mail from Lynn Rogut, then deputy director of the Robert Wood Johnson Foundation Investigator Awards in Health Policy Research program. Lynn's e-mail suggested that I apply for one of the program's very generous grants, which I very quickly did. My proposal laid out the basis of this book, and becoming a recipient of an Investigator Award in Health

Policy Research from RWJF made it possible. To all those at the RWJF program, I am deeply grateful, and particularly to David Mechanic, Lynn Rogut and Cynthia Church at Rutgers, who oversaw the program during the three years of my grant. I am also indebted to the University of California, Berkeley, and the late (and very much missed) Pat Buffler, and her colleagues Amber Sanchez and Theresa Saunders at the School of Public Health, who administered the grant and gave me an academic base for my research.

Chapter 8, "Defending Sugar," began its existence as the article "Sweet Little Lies" in the November/December 2012 issue of *Mother Jones.* That article was a joint venture with Cristin Kearns, who first introduced herself to me in February 2011, after a talk I gave at a Denver independent bookstore. Cristin was then a working dentist, but she told me how she had taken it upon herself to investigate the sugar industry and had discovered a cache of confidential Sugar Association, Inc., documents exposing its public-relations strategy in the 1970s. Those documents became the basis of the *Mother Jones* article and now Chapter 8 as well. Cristin's investigative skills, writing, and critical thinking were indispensable to both. (The story can be read online at: http://www.motherjones.com/environment/2012/10/former-dentist-sugar-industry-lies.) I also have to thank the staff at *Mother Jones* who shepherded the article through to publication — particularly Mike Mechanic (son of David), Maya Dusenberry, Maddie Oatman, Elizabeth Gettleman, and Cathy Rodgers.

The arguments that ultimately constitute the case against sugar had a first public run in April 2011 in a *New York Times Magazine* cover article entitled "Is Sugar Toxic?" I'm grateful to Hugo Lindgren, Vera Titunik, David Ferguson, and the magazine's staff (circa 2011) for helping me make those arguments fit for public consumption.

I'd like to thank Clarke Read and Maya Dusenberry (again) for their extensive help with research for this book, and Nathan Riley, Devon Simpson, and Ethan Litman, who also contributed their research skills. I'm grateful to Dan Palenchar and my old and dear friend Scott Schneid for doing what they could to help me get the facts straight. Mark Friedman, Michael Rosenbaum, and Robert Kaplan took the time to read this book in draft and help me get my thoughts straight, and for that I'm equally grateful. Any errors that remain, of course, are my responsibility alone. I'd like to thank Jeffrey Mifflin, archivist at Massachusetts General Hospital in Boston, and Stacey Peeples, curator–lead archivist at Pennsylvania Hospital in Philadelphia, for their generous assistance in providing diabetes inpatient data from their hospitals going well back into the nineteenth century.

I'm grateful to my agent, Kris Dahl at ICM, for what is now three decades of unwavering support. And I couldn't be more beholden to my editor, Jonathan Segal at Knopf, who has supported my nutrition writing from the beginning and supported me as a writer in the process. He's the kind of editor that every writer dreams of having. I'd also like to thank, at Knopf, the editorial assistant Julia Ringo, publicist Jordan Rodman, production manager Claire Ong, and text designer Maggie Hinders. And special thanks to production editor Victoria Pearson.

All three of my books on nutrition and chronic disease are pleas ultimately for better nutrition science, and for the rigorous trials necessary to test critical assumptions about a healthful diet that have been publicly embraced over the years as dogma. Laura and John Arnold and their colleagues at the Laura and John Arnold Foundation have embraced this belief that better and more critical nutrition research is necessary for the health of the nation, and have been willing to act upon it philanthropically. For that I will always be grateful. I'd also like to thank all my colleagues over the years at the Nutrition Science Initiative for their support and friendship, and for making it possible to fund and facilitate what we consider the first stage of necessary studies.

If my bias against sugar isn't blindingly clear by now, then I'll establish it beyond doubt by saying that I am deeply grateful as well to those researchers and physicians who had the temerity to take a stand against sugar, knowing that at least some proportion of their professional colleagues would criticize them for doing so. Peter Cleave and John Yudkin played critical roles, as I discuss in the book, and should be thanked by all for doing so. Robert Lustig at the University of California, San Francisco, has recently taken up Yudkin's torch and been singularly effective at forcing the public and scientific discussions on sugar and health. Richard Johnson at the University of Colorado continues to do unique, and what may be vitally important, research, and I fear I didn't give it nearly the treatment and discussion it deserves. For narrative reasons,

William Dufty's contribution to this ever-evolving controversy — the massively best-selling *Sugar Blues,* first published in 1975 — is not mentioned in these pages, but he has to be acknowledged and thanked nonetheless. I would also like to thank and acknowledge Connie Bennett, Nancy Appleton, Ann Louise Gittleman, and the numerous other nutritionists, dietitians, and physician-authors who have publicly taken up this cause.

Finally, my wife, Sloane Tanen, has ultimately made this book possible with her love and support and her humor, not to mention her cheerful willingness, weekend after weekend, year after year, to take our boys to friends' houses and sporting events (while occasionally humming "Cat's in the Cradle") as their father withdrew to his office once again to work on a book or tilt at a windmill. To those boys, Nick and Harry, as always, go my eternal thanks, for reminding me why I do this, and keeping their sense of humor in the process.

NOTES

Epigraphs. "We are, beyond question": Anon. 1857.
"I am not prepared": Chaudhuri and Esterl 2016.

Author's Note

A third of all adults: CDC 2016b.
One in seven is diabetic: Menke et al. 2015.
Die of cancer: ACS 2016.

Introduction: Why Diabetes?

Epigraph. "Mary H— an unmarried woman": Quoted in Feudtner 2003: 45.
The patient was Mary Higgins: Ibid.: 45–48. See also Wright 1990: 325.
"hundreds of volumes": Fitz and Joslin 1898.
"wholesome tendency": Ibid.
Joslin published an article: Joslin 1921.

Fifty years ago . . . today: NIDDK 2012.
Double worldwide since: WHO 2015.
Back . . . to the nineteenth century: Helmchen and Henderson 2004.
Sushruta, a Hindu physician: Tattersall 2009: 10.
"The patient does not survive": Aretaeus 1837: 1–3.
"observ[ing] an extensive range": Rollo 1798.
Mortality records from Philadelphia: Vaughan 1818.
Footnote. E-mail, Jeffrey Mifflin, archivist, Massachusetts General Hospital, Jan. 15, 2014.
"rarer diseases . . . seven years' practice . . . The truth": Saundby 1891: 1, 26, 34.
Of the thirty-five thousand patients: Osler 1892: 296.
Next eight years: Osler 1901: 418.
Mortality statistics: Osler 1909: 409.
an epidemic of diabetes: Yearly diabetes admissions at Pennsylvania Hospital were provided by Stacey Peeples, curator–lead archivist, Pennsylvania Hospital, in an e-mail on March 12, 2009.
In Copenhagen: Joslin 1934.
400 percent increase: Emerson and Larimore 1924.

Rapidly becoming a common disease: Joslin 1934.
One in every seven to eight: Menke et al. 2015.
Another *30 percent:* Gregg et al. 2014.
Almost two million Americans: CDC 2014b.
Patients admitted to VA hospitals: VHA 2011.
Die at greatly increased rates: ADA 2014.
A dozen *classes:* Khardori 2015.
Thirty billion dollars: ADA 2013.
"Diabetes is in all cases": Saundby 1901.
"The incidence of diabetic morbidity": Wilder 1940: 38.
"appalling increase": Joslin 1950.
"one of the most important human problems": West 1978: ix.
China at the turn of the twentieth century: Saundby 1908; Reed 1916.
In the 1980s . . . the latest estimates: Xu et al. 2013.
Among Inuit through the 1960s: Sagild et al. 1966; Schaefer 1968.
"Eight Alaskan Eskimos": Mouratoff et al. 1967.
By the 1970s: Mouratoff and Scott 1973.
In recent studies: Jørgensen et al. 2012.
In Native American tribes and First Nations Peoples: Young et al. 2000.

Sandy Lake: Abraham 2011.
Data in Native American populations: West 1974.
Navajo from the 1950s: Sugarman et al. 1990.
Similar patterns: West 1978; Zimmet et al. 2001; IDF 2015.
"Some had been nomadic hunters": West 1974.
"Rises and falls": Emerson and Larimore 1924.
The sugar industry hired pollsters: National Analysts 1974: 33.
"You need an insulin shot": Bruce and Crawford 1995: 213.
"plays an etiological role": McGandy and Mayer 1973.
Researchers and clinicians: See, for instance, NAS 1975.
"The fundamental cause": WHO 2015.
"we eat too damn much": *Today* show 1976.
Attempts to prevent diabetes: See, for instance, DePue et al. 2010; Mau et al. 2010.
An 800 percent increase: CDC 2014a.
Healthy Weight Commitment Foundation: Starling 2009.
"not about demonizing": *PBS NewsHour* 2010.
One in four Americans: NIDDK 2014b.

A conservative estimate: The CDC estimates the direct and indirect costs for heart disease and stroke at $315 billion each year, cancer at $157 billion, diabetes at $245 billion, and obesity (in 2008) at $147 billion (CDC 2016a). The Rand Corporation has estimated the total monetary cost of dementia, including Alzheimer's, at between $157 and $215 billion (Hurd et al. 2013).

Alzheimer's as type 3 diabetes: See, for instance, Guthrie 2007.

"We are to admit": See https://en.wikiquote.org/wiki/Isaac_Newton.

"Everything should be": See https://en.wikiquote.org/wiki/Albert_Einstein.

"multifactorial, complex disorders" or "multidimensional diseases": See, for instance, NIDDK 2011: 117–38.

At least a tenth of all cases of lung cancer: ALA 2014: 5.

"all wars combined": West 1978: ix.

Heavy smokers had twenty to thirty times: See, for instance, Doll and Hill 1964.

This confusion still exists: See, for instance, Reynolds 2014; Seidenberg 2015.

Footnote. Ventura et al. 2011.

HFCS was the cause: See, for instance, Bray et al. 2004; Pollan 2002.

"the flashpoint": Interview, Marion Nestle,

Jan. 5, 2011.

Corn Refiners Association petitioned: Wells 2014.

FDA denied the Corn Refiners' petition: Landa 2012.

"not the single hint": Tappy and Lê 2010.

Per capita consumption numbers cited by government: See, for instance, Putnam and Haley 2003. USDA

reports that 114 pounds of sugar: See Table 49 and Table 50 at http://www.ers.usda.gov/data-products/sugar-and-sweeteners-yearbook-tables

Food and Drug Administration report: Glinsmann et al. 1986.

"Limitations on accurately": USDA 2016.

Americans consumed *only:* See Table 51 and Table 52 at http://www.ers.usda.gov/data-products/sugar-and-sweeteners-yearbook-tables.

"We perceive it": quoted in Strom 2012.

Chapter 1: Drug or Food?

Epigraphs. "The sweet shop": Dahl 1984: 33.

"Imagine a moment": Pollan 2001: 18.

"a near invulnerability": Mintz 1985: 99.

"an innocent moment": Richardson 2002: 292–93.

Sugar-intolerant Canadian Inuit: Ellestad-Sayad et al. 1978.

Equivalent of 360 eggs: Deerr 1950: 529.

"The depression proved": Ripperger 1934.

"whether [sugar] is actually": Mann 2011: 289.

"That sugars, particularly": Mintz 1985: 100.

"drug foods": Ibid.: 99.

Sugar was used to sweeten liquors: Courtwright 2001: 29.

"sublimated essence": Quoted in Pendergrast 1993: 194.

Single most widely distributed: Ibid.: 439.

Morphine addiction and "Like Coca": Quoted in ibid.: 24–25.

"marriage of tobacco and sugar": Weiss 1950: 2.

"eighteenth-century equivalent": Ferguson 2002: 13.

"some compensation": Mann 2011: 372.

"In nutritional terms": Barker et al. 1970.

"an ideal substance": Mintz 1985: 186.

"the perfect pleasure": Wilde 1908: 106.

"greedily suck down": Slare 1715: 8.

"a marked relaxation": Steiner 1977.

Why humans evolved a sweet tooth: See, for instance, Bramen 2010.

The world's greatest sugar consumers . . . lacked any succulent fruit: Mintz 1991.

holds for Australians: Anon. 1928b.
"distress vocalizations": Blass 1987.
Sugar will allow adults: Gardner 1901.
More effective than breast milk: Ors et al. 1999.
Cats don't, for instance: Kare 1975.
Cattle, on the other hand: Anon. 1886.
Agronomists reported: Plice 1952.
"false palatableness": Anon. 1884.
The actual research literature: See, for instance, Avena et al. 2008; Schmidt 2015.
Serge Ahmed has reported: Ahmed 2012.
"There is little doubt": Quoted in Anon. 1909.
The Big Book: AA 2001: 133–34.
"tremendous increase": Anon. 1919a.
"wreckage of the liquor business": Anon. 1920.
"due to prohibition . . . never heard of a man": Anon. 1925b.
From the 1600s . . . quadrupled again: Deerr 1950: 490–91, 532.
Sixteen-fold: Woloson 2002: 187.
"development of the sugar appetite": Anon. 1909.
too damn much: *Today* show 1976.

Chapter 2: The First Ten Thousand Years

Epigraph. "M. Delacroix": Brillat-Savarin 1986: 104.

"English Man's Fly": Warner 2011: 169–70.

Native Americans using maple syrup: Root and de Rochemont 1976: 40–41.

"yields a sugar": Warner 2011: 162.

Sorghum . . . had a run: Galloway 1989: 2–3.

"kindled an enthusiasm": Warner 2011: 147.

Anthropologists believe: On the history of sugar and sugarcane, see, for instance, Prinsen Geerligs 2010; Deerr 1949; Deerr 1950; Aykroyd 1967; Mintz 1985; Richardson 2002 (17 percent sugar: 69); Abbott 2007.

Creation myths: Cohen 2013.

"a series of liquid-solid operations": Mintz 1985: 22.

Would state in its defense: See, for instance, Stare 1976b.

The only pure chemical substance: Mintz 1985: 22.

Sugar is extraordinarily useful: Pennington and Baker 1990.

"I must remove": Deerr 1949: 68.

A thousand pounds . . . Ramadan feasts: Ibid.: 92.

"sucking enthusiastically": Mintz 1985: 28.
"a most precious product": Phillips 1985: 93.
Italian city-states: Prinsen Geerligs 2010.
Kitchen expenditures of Henry II and Edward I: Mintz 1985: 82.
"to no avail": Aykroyd 1967: 26.
"eaten with the end in mind": Mintz 1985: 99.
"good for almost every part": Walvin 1997: 99.
"who kept it exclusively . . . out of gluttony . . . No food refuses": Montanari 1994: 120–21.
"Sugar spoils no dish": Braudel 1992: 191.
"swinging machetes": Mann 2011: 139.
Sugar and slavery went hand in hand: Because the relationship was so intimate, this history is told at length in both the histories of sugar and the histories of slavery. Of particular use to me was Phillips 1985.
Columbus who first brought sugar: Deerr 1949: 115–23.
Portuguese colonists in Brazil: Ibid.: 104.
"the whole of Christendom": Ibid.: 138.
"better to be tossed out": Huetz de Lemps 1999: 385.
The British: Deerr 1949 (Jamestown: 148; Barbados and Jamaica: 158–66; number

of slaves on Barbados: 166; *footnote:* 106–8).

Twelve and a half million Africans: This estimate is from slavevoyages.org, and considered the most authoritative estimate available.

A fifth of all British imports: Ferguson 2002: 61.

"second addiction": Proctor 2011: 49.

Sugar was an ideal target of taxation: For this and the history of taxation, see Mintz 1985: 188–95; Strong 1954: 87–107.

"from which they could obtain slaves": Burrows and Wallace 1999: 72.

By 1810 . . . 1860: Deerr 1950: 462.

"without which the West Indian plantations": Burrows and Wallace 1999: 120.

"molasses was an essential ingredient": Mintz 1991.

"luxury of kings": Mintz 1985: 96.

"the delight of childhood": Anon. 1873.

More than half a billion: Moore 1890.

Development of the beet-sugar industry: Deerr 1950 ("To scientific ability": 475; "into the Thames": 478).

More than 15 percent: Woloson 2002: 31.

U.S. Department of Agriculture: Warner 2011 ("be numbered": 91).

Footnote. Ibid.: 19.

By the 1920s: This comparison is based on Anon.

In the 1820s, 1921a, assuming Deerr's statistics (Deerr 1950: 462) that at least ten refineries were operating in New York City.

The manner in which we consumed it: Mintz 1985: 129–47.

Mark Twain wrote of his youth: Twain 2010: 2.

Sugar was added by the bakers: Hess and Hess 2000: 57–60.

Sugar content greater than 10 percent: Pennington and Baker 1990: 132.

Candy: Woloson 2002: 33–40 ("display of grown-up prestige . . . a venue for the children": 33).

Centennial Exposition: Richardson 2002: 327.

By 1903: Anon. 1903.

Chocolate: Woloson 2002: 144–50.

Chocolate staples: CandyFavorites.com, at http://www.candyfavorites.com/shop/history-american-candy.php.

Ice cream: Quinzio 2009: 75–102.

"not only a new treat": Woloson 2002: 88.

Ice-cream sundae: This and the other inventions are in Quinzio 2009: 127, 173, 174, 175.

Footnote. Ernest Hamwi: Quinzio 2009:

159; Pendergrast 1993: 13.

Coca-Cola: This history comes primarily from Pendergrast 1993 ("the magnificent competitors": 463; "I don't know anything": 89; and "a valuable Brain Tonic": 29).

Pepsi-Cola came along: Stoddard 1997 (syrup sales increased: 26–28).

Cuban and American industries: Babst 1940: 57–59.

"The people of Europe": Anon. 1921b.

Three billion bottles: Anon. 1919b.

Chapter 3: The Marriage of Tobacco and Sugar

Epigraph. "Such an investigation": Weiss 1950: 2.

Lung cancer: For figures on annual deaths from lung cancer, see Proctor 2011: 57.

"This business of sugar": Proctor 2011: 33.

"Were it not for sugar": Weiss 1950: 2.

Camel was the best-selling cigarette: Ibid.: 6.

"When the smoke is inhaled": Garner 1946: 436.

Could have made cigarettes that were harder to inhale: Proctor 2011: 34.

"flue-curing may well be": Proctor 2011: 34.

“the closest parallel”: Weiss 1950: 18.
German researchers noted: Proctor 2011: 34.
“objectionable properties”: Garner 1946: 442.
“candied up”: Proctor 2011: 31.
Footnote: Tilley 1972: 512.
“Sugar enhances”: Weiss 1950: 31.
“the perverted tastes”: Ibid.: 514.
Act of “necessity”: Weiss 1950: 5.
Fifty million pounds of sugar: Tilley 1972: 622–23.
Footnote: Weiss 1950: 39.
“This [caramelization] process”: Ibid.: 45.
“Consumer acceptance”: Talhout et al. 2006.
“acid buffering capacity”: Elson et al. 1972.
“This spectacular development”: Weiss 1950: 64–65.

Chapter 4: A Peculiar Evil

“In hard times”: Courtwright 2001: 98.
“The peculiar evil”: Orwell 1958: 32.
“depression-proof”: See, for instance, Krauss 1947 on the soft-drink industry.
Sixteen pounds *higher:* Ripperger 1934.
Coca-Cola thrived, as did Pepsi: Pendergrast 1993 (225 percent: 174; “breakfasting on Coca-Cola”: 174).

“price inelastic”: Marks and Maskus 1993.
Cycles invariably begin: Borrell and Duncan 1993; Hannah and Spence 1996: 46–67.
“frantic and abnormal”: Babst 1940: 23.
“the unhealthy economics”: Anon. 1945a.
China, for instance: Anon. 1931.
Sugar Act: Schmitz and Christian 1993; Walter 1974; Babst 1940.
“the most powerful”: Belair 1937.
“benefit payments”: Swift 1937.
By 1935: Quinzio 2009: 177.
Coca-Cola and Pepsi: Pendergrast 1993: 176–77.
Sales nearly quadrupled: Krauss 1947.
Seventy pounds: White 1945.
“worst sugar famine”: Williams 1945.
“It would not seem unreasonable”: White 1945.
“our warriors”: Flanagan 1943.
A hundred million pounds: Anon. 1944b.
“underestimate the importance”: Anon. 1944a.
“fighting food value . . . to correct popular misinformation”: Anon. 1944b.
Pepsi circumvented: Stoddard 1997: 95–98.
Coca-Cola: Pendergrast 1993 (“friends and customers . . . sampling and expansion”: 212; *footnote:* 210; “serve those two bil-

lion . . . When we think of Communism": 236).
Coca-Cola on the cover in 1950: Ibid.: 232.
Pepsi quickly catching up: Stoddard 1997: 12–131.
Nixon with Khrushchev: Pendergrast 1997: 269.
Ice-cream consumption alone doubled: Quinzio 2009: 200.
Canned breakfast juices: Hamilton 2009.
"crowning achievement . . . perhaps a defining moment": Lovegren 2012: 213.
Gallons of fruit juice a year: ERS 2015.
Breakfast cereals: Bruce and Crawford 1995.
Kellogg and Post: Ibid.: 10–59 ("The causes of indigestion": 17).
"he felt that sugar": Ibid.: 50–51.
"America's sweet tooth": Ibid.: 214.
"Sickened by the sugary excess": Ibid.: 103.
"turn into bricks": Ibid.: 106.
Post then began the trend: Ibid. ("trading off sugar carbohydrates . . . the nutritional value": 106; "a charitable organization": 108).
Kellogg's set out: Ibid. ("it was their salvation": 109; "all this sweetness . . . a dietary flop": 111).
"possible dietary effects": Ibid.: 111.
six hundred million dollars: Ibid.: 240.

Candylike nature: Ibid. ("It tastes like maple sugar": 158; "like a chocolate milk shake": 155; "Eating any of the cereals": 261).

Chapter 5: The Early (Bad) Science

Epigraphs. "In spite of the doctors": Anon. 1856.

"Most people know": Willaman 1928.

Blaming sugar for a host of ills: See, for instance, Emerson and Larimore 1924 (diabetes); Thorne 1914 (cancer); Dix 1904 (rheumatism); Anon. 1909 (gallstones, jaundice, liver disease, inflammation, gaseous indigestion and sleeplessness); Anon. 1928a (ulcers and intestinal diseases); Lawrie 1928 ("nervous instability"); Anon. 1910 ("a degenerate people").

"No other element": Gibson 1917.

Science of nutrition: On the history of nutrition and the roots of modern nutrition, see, for instance, Lusk 1933; Rose 1929.

"The amount of information": Atwater 1888.

The radioimmunoassay and the modern era of endocrinology: Karolinska Institute 1977.

Medicine and science had little connection:

See Flexner 1910; Ludmerer 1988 (Bowditch: 37); Shryock 1979; Rosenberg 1987.

"Scientists are not so much": Krebs 1967.

"promotes nutrition *and* [my italics] corpulency": Deerr 1949: 46.

"The pissing evil": Willis 1679.

"wonderfully sweet . . . an ill manner of living": Ibid.

Footnote. "We meet with examples": Willis 1679.

An exaggeration: Robert Tattersall, personal e-mail, July 1, 2013.

"disapprove[d] [of] things": Willis 1685: 372.

"frighten the Credulous": Slare 1715: 22.

To "defraud" infants: Ibid.: 8.

"near Sixty-seven . . . . I write without Spectacles . . . were bitter enemies": Ibid.: 63.

Footnote. "That which preserves": Ibid.: 59.

"the worst of the Skum": Ibid.: 19.

Less than five pounds per capita: Hannah and Spence 1996: 10.

"fine proportions . . . inclining to be too fat . . . so very high a Nourisher": Ibid.: E4.

"scarcely anything": Moseley 1799: 157; "Give a negro infant . . . old, scabby, wasted . . .": Ibid.: 144.

Abel Jordão suggested: Jordão's lectures and article were summarized in two reviews in *The American Journal of Medicine:* Jordão 1866; Jordão 1867 ("a robust adipose constitution").

"On this same principle": Brigham 1868.

"without any waste": Gardner 1901.

"nutritive value": Higgins 1916.

"unexpected stimulating properties": Gardner 1901.

Parisian cab companies: Ibid.

"At great elevations": Anon. 1926.

"The results were conclusively": Gardner 1901.

"sugar training . . . did not become 'stale' ": Ibid.

"pound of peppermints . . . preposterous": Anon. 1926.

Footnote. Anon. 1924.

"an overdose of insulin": Kohn et al. 1925.

"The most curious thing": Anon. 1925a.

"would seem to be a food": Abel 1915: 30.

"the popular prejudice": Gardner 1901.

"one of the most valuable articles": Gardner 1901.

"it may with benefit": Anon. 1887.

"growing world-wide abstinence": Anon. 1929.

"splendid alternative": Proctor 2011: 61.

"The consumption of sugar": Allen 1913: 146.

"large quantities": Ibid.: 148–49.

"open to accusations against sugar": Ibid.: 146.

"diabetes in the tropics": Charles 1907.

"not the slightest shadow": Allen 1913: 147.

"Unless the unknown cause": Ibid.: 147–48.

"If he is a poor laborer": Ibid.: 152.

Metropolitan Life: Anon. 1923.

New York State: Emerson and Larimore 1924.

His textbook: Joslin 1916.

Footnote. Kahn et al. 2005.

"No child can grow up": Feudtner 2003: 133.

The value of sugar for athletes: Anon. 1925d.

Footnote. Anon. 1925d.

"An orange is less temptation": Joslin 1923: 74.

"Indeed, a high percentage": Joslin 1917: 59.

Footnote. Snapper 1960: 374.

Blamed diabetes on the automobile: Anon. 1925c.

"an excess of fat": Joslin 1927.

"While there is a popular conception": Long 1927.

Diet relatively rich in carbohydrates: Himsworth 1931b ("Sugar is what must be given"); Himsworth 1931a.

Himsworth would later report: Himsworth 1949a (diabetes rates had risen); Himsworth 1949b ("It would thus appear").

Footnote. Himsworth 1935. Inuit on Baffin Island: Heinbecker 1928.

"Fisherfolk": Mitchell 1930.

Joslin would describe . . . Himsworth in turn: See, for instance, White and Joslin 1959 ("painstakingly accumulated": 70); Himsworth 1935; Joslin 1934; Mills 1930; Joslin 1928: 165.

As late as 1963: Insull et al. 1968.

Himsworth himself rejected it: Himsworth 1949a.

Subject of whether or not sugar consumption: Marble et al., eds., 1971.

Chapter 6: The Gift That Keeps on Giving

Epigraphs. "Diabetes . . . is largely a penalty": Joslin 1921.

"18 CALORIES!": Bart 1962.

No profound revelations to be gleaned: See, for instance, FAO n.d.

"Which is LESS FATTENING?": Domino Sugar 1953.

"the ingestion of a quantity": von Noorden

1907: 693.

Louis Newburgh: Newburgh and Johnston 1930a ("All obese persons . . . perverted appetite . . . lessened outflow"); Newburgh and Johnston 1930b ("various human weaknesses").

"the whole problem of weight": Anon. 1939.

Footnote. "To attribute obesity": Mayer 1968: 7.

"That which the body needs": von Bergmann and Stroebe 1927.

Bauer confirmed the obvious: Bauer 1929.

"equivalent to that of height": Friedman 2004.

"a good or poor appetite": Newburgh 1942. Joslin, apparently, believed the same: Wilder and Wilbur 1938: 312.

Bauer had spent his professional career: Anon. 1979.

"The genes responsible": Bauer 1940. (The best source in English for Bauer's observations on obesity is Bauer 1941.)

this "well known phenomenon": Stockard 1929.

"Probably she does not know": Newburgh 1942.

"The energy conception": Grafe 1933: 148.

Bauer took up Bergmann's thinking: Silver and Bauer 1931; Bauer 1940; Bauer 1941 ("a malignant tumor . . . a sort of anarchy").

"deserves attentive consideration": Wilder and Wilbur 1938: 312.

"more or less fully accepted": Rony 1940: 173–74.

The primary German textbook: Bahner 1955.

"The work of Newburgh . . . Newburgh answered that": Anon. 1955c.

Animal models: See, for instance, Lee and Schaffer 1934; Hetherington and Ranson 1939; Hetherington and Ranson 1942; Brooks 1946; Brooks and Lambert 1946; Mayer 1953b; Alonso and Maren 1955; Levitsky et al. 1976; Mrosovsky 1976; Greenwood et al. 1981: Oscai et al. 1984 (high-fat diets); Sclafani 1987 (high-sugar diets); Cohen et al. 2002; Bluher et al. 2003.

"is also probably present": Cahill 1978.

It was the invention of Rosalyn Yalow: Yalow and Berson 1960.

"a revolution": Karolinska Institute 1977.

Answers began coming: Berson and Yalow 1965.

"the negative stimulus . . . lipogenic": Ibid.

A second revelation: Ibid.

Falta and Himsworth: For a good review of

their work on insulin resistance, see Gale 2013.
"We generally accept": Berson and Yalow 1965.
By assuming that hyperinsulinemia and insulin resistance: See, for instance, NIDDK 2014a.
"It is a medical fact": Borders 1965.
"knock down reports": Anon. 1956.
"that are spent as energy . . . Sugar is neither": Sugar Information, Inc., 1956.
"Sugar Bowled Over": Anon. 1955b.
"shift blame for obesity": O'Connor 2015.
"fringe view": Snowden 2015.
"champions of energy balance": GEBN 2015b.
The GEBN Web site noted: GEBN 2015a.

Chapter 7: Big Sugar

Epigraph. "If . . . every American": Anon. 1955a.
"cut-throat competition": Barnard 1928.
To build up the immune system: Sugar Institute 1931b.
Enhancement of iced beverages: Sugar Institute 1931a.
"Recent scientific investigations": Sugar Institute 1930.
"repressive methods": Anon. 1932.

Supreme Court, which ruled: Anon. 1936b.

Sugar Institute was dissolved: Anon. 1936a.

Surprising number of Americans: Levenstein 1993: 53–68 ("of all foods": 68).

"For Health . . .": at https://research.archives.gov/id/514288.

"food faddists . . . sugarcoating the bitter . . . a heavy barrage . . . HOW MUCH SUGAR": Sugar industry document: Lamborn 1942.

Council on Foods and Nutrition report: CFN 1942.

"Don't complain": Anon. 1942a.

"Coffee without sugar today": Lamborn 1942.

"A suggested program": Ibid.

Three million dollars in research: Anon. 1951a.

SRF/SAI grants went to . . . prominent researchers: Anon. 1945b.

First award went to MIT: Anon. 1943.

President of MIT would later say: Anon. 1942b.

Among the many other researchers: See, for instance, Hockett 1947.

Footnote. Hockett: Sourcewatch, at http://www.sourcewatch.org/index.php/Robert_Casad_Hockett.

Cavities and tooth decay had been linked to sugar: Aykroyd 1967: 117–26; Mintz 1985 ("a defect the English": 134; "rotteth the teeth": 105).

Then it began to explode: Suddick and Harris 1990.

"startlingly high proportion": Drummond and Wilbraham 1994: 387.

"You would have to look": Orwell 1958: 33.

Price . . . published seminal study: Price 1939.

"dental caries was not": Fosdick 1952.

"a nice place to live": Ibid.

By the 1930s: See, for instance, Anon. 1934.

University of Iowa and Harvard: Anon. 1945b.

By 1950: Kearns et al. 2015.

According to the SAI's annual report: Kearns et al. 2015.

"most of the present counsel": Smith 1952.

"stands little chance": Anon. 1951a.

"prompt brushing after every meal": Smith 1952.

"Millions of Americans . . . America's No. 1 health problem": Anon. 1953.

"the great American dieting neurosis": Walker 1959.

Cases of "low-calorie" soft drinks had been sold: Walker 1959.

American Sugar Refining Company . . . campaign: Anon. 1951b.

Sugar Association took over: Anon. 1954.

Footnote. Ewen 1998 ("Sultan of Sell").

Physicians at Harvard: Williams et al. 1948.

Cornell: Reader et al. 1952.

Stanford: Cutting 1943.

The occasional medical textbook: Greene, ed., 1951: 348.

"neither a 'reducing food' ": Sugar Information, Inc., 1956.

Idea of Jean Mayer: Mayer 1953a.

Funded by . . . the Sugar Association: Cheek, ed., 1974: 100–103.

Refuted in experiments: See, for instance, Bernstein and Grossman 1956.

"satisfies the appetite faster . . . takes the edge off": Sugar Information, Inc., 1956.

"Q. How can sugar help": Sugar Information, Inc., 1957.

This competitive advantage: See House Committee 1970: 6; Cray 1969.

Saccharin had been discovered: Priebe and Kauffman 1980; Cohen 2006 ("first time in history": 96); Warner 2011: 181–207.

Roosevelt . . . argument . . . with Wiley: Cohen 2006: 96–7 ("thought he was eating sugar").

He had begun his career: Warner 2011: 92–93.

"anybody who says": Cohen 2006.

"bearing out an old aphorism": Handler 1975.

Not how the FDA saw it: Warner 2011: 187–89.
Cyclamates did not have: Ibid.: 195–207.
FDA required the same labeling: Ibid.: 197.
Coke and Pepsi released: Nagle 1963.
Began doubling yearly: Nuccio 1964.
Analysts were predicting: Nagle 1965.
Sugar industry responded: Anon. 1964.
"If it's wrong": Ibid.
Ways to diversify their products: Frost 1965.
None of these held the promise: Hickson 1975: 24–25.
"find new arguments": Hickson 1962 "If anyone can undersell": Cray 1969.
"Delaney clause": U.S. Congress 1958 amendment ("No additive shall": 1786).
Between 1963 and 1969: Kelly 1969.
FDA published . . . concluded that there was little to fear: Warner 2011: 200.
WARF researchers would publish: Nees and Derse 1965.
"mental disturbance": House Committee 1970: 23.
"had an understandable interest": House Committee 1970: 23–24.
Researchers funded by Abbott Laboratories: Warner 2011: 201–2.
"you'd drown before": Pendergrast 1993: 290.
FDA administrators had originally hoped:

Warner 2011: 202; House Committee 1970: 24.

"one of its primary missions": NAS 1975: 219.

"supreme scientific politician": DGF 1972.

"in excess of the amount": Lyons 1977.

"It's humanly impossible": Rhein and Marion 1977: 58.

FDA succumbed to . . . a warning label: Priebe and Kauffman 1980; Warner 2011: 203–4.

Considers neither cyclamates nor saccharin to be carcinogenic: NCI 2009.

Surge in diet-soda sales that failed to last: Timberlake 1983; Anon. 2016; interview, Manny Goldman, consumer products consultant, March 21, 2002.

Chapter 8: Defending Sugar

Epigraphs. "If we are looking": Yudkin 1963. "So the real question": NAS 1975: 96.

Tatem spoke . . . to Chicago Nutrition Association: Tatem 1976c ("purest and most economical . . . opportunists dedicated . . . promoters and quacks . . . calculatedly enlist . . . neatly apply . . . wade through").

In Scottsdale, Arizona: Tatem 1976a ("enemies of sugar . . . persuasive purveyors . . . successfully misled . . . "sugar, once ac-

cepted").
"the limited bill": Mayer 1976.
"scientific farce": Tatem 1976c.
"We have moved to the defensive": Ibid.
"common food ingredients": USFDA 1958.
"one of the offshoots . . . establish the facts": Tatem 1976c.
In 1948, the American Heart Association: Anon. 1948a; Anon. 1948b; Davies 1950; Moore 1983: 77.
Russian researchers had famously: Anitschkow and Chalatow 1913.
Keys had a conflict of interest: SRF 1945: 16.
Combative and ruthless: See, for instance, Blackburn n.d.
"uncompromising stands": Page et al. 1957.
"best scientific evidence": AHA 1961.
Keys was on the cover of *Time:* Anon. 1961.
Suggested that eating less saturated fat would shorten our lives: Frantz et al. 1989.
"suggestive" evidence: Hooper et al. 2015.
"infants, children, adolescents": Inter-Society Commission 1970.
Footnote. "We never saw the results": Interview, I. D. Frantz, Jr., Dec. 9, 2003.
"an unproved hypothesis": Dawber 1978.
Fat consumption may have increased: Taubes 2007: 10–13.
"We now eat in two weeks": Yudkin 1963.

Cohen had spent the previous decade: Cohen 1963.

"The quantity of sugar": Cohen et al. 1961.

"absolutely staggered by the difference": Campbell's testimony in Select Committee 1973: 208–18.

Campbell focused his research: Campbell 1963; Cleave and Campbell 1966 ("a veritable explosion . . . almost certainly": 25).

Footnote. Ibid.

"a *starvation wage* . . . enormously fat": Select Committee 1973: 213.

Urban and rural Zulu populations: Campbell 1963 ("a remarkably constant period").

Cleave was an outsider: On his background, see Wellcome Library, "Cleave, 'Peter' (1906–1983)." At http://www.aim25.ac.uk/cgi-bin/search2?coll_id=4602&inst_id=20.

Cleave had been arguing . . . since 1940: Cleave 1940.

"Law of Adaptation": Cleave and Campbell 1966 ("an adequate period": 1).

"Such processes": Cleave 1956.

"A person can take down": Cleave 1975: 8.

"Assume that what strains": Ibid.: 84.

Jacques Monod would later credit: Monod 1965.

"anatomically, physiologically, and biochemically": Yudkin 1963.

Attention away from cholesterol: See, for instance, Sniderman et al. 2011.

Yale and Rockefeller researchers: Albrink et al. 1962; Albrink 1963; Albrink 1965.

Rockefeller researchers were reporting: Ahrens 1957; Ahrens, Hirsch, et al. 1957; Ahrens, Insull, et al. 1957; Ahrens et al. 1961.

Yudkin tested his sugar hypothesis: See, for instance, Szanto and Yudkin 1969; Yudkin et al. 1969; Bender et al. 1972; Yudkin 1986: 94–103.

Footnote. Anderson et al. 1963; Grande et al. 1974.

Cardiologists and the American Heart Association thought: See, for instance, Anon. 1989.

"The refining of sugar": Dickson 1964.

Sugar Association first became concerned: Hickson 1962.

"Castro Situation": Hass 1960.

"top priority": Kelly 1969.

"What's at Stake in Sugar Research": Kelly 1969.

"educating health professionals": Sugar Association, Inc., at http://www.sugar.org/about-us/.

Yudkin had implicitly attacked Keys: Yudkin 1957.

Keys returned the favor: Keys 1971.

"adequate to explain": Ibid.

"alone in his contentions": Keys and Keys 1975: 58.

"quite a bit of loathing": Interview, Richard Bruckerdorfer, Feb. 12, 2004.

During the Korean War: See, for instance, Mayer and Goldberg 1986; Enos et al. 1953.

French traditionally consumed far less sugar: Huetz de Lemps 1999.

"Sweetness does not seem": Mintz 1985: 190.

"does not have widespread support": Brody 1977.

"Although there is strong evidence": Masironi 1970.

Truswell, who believed and argued publicly: Truswell 1977.

Ended his research career: Interviews, Richard Ahrens, Dec. 7, 2002; Donald Naismith, Dec. 11, 2002; Richard Bruckendorfer, Jan. 29, 2003, and Feb. 12, 2004; and Michael Yudkin, Feb. 13, 2004.

Popular polemic against sugar: Yudkin 1972a; Yudkin 1972b.

"Sugar — The Question Is": Warren 1972.

A Senate subcommittee: Select Committee 1973.

The testimony came from: Select Committee 1973 (“The only question”: 256; “and they die”: 155).

“The research and findings”: Hillebrand, ed., 1974: 56.

“From the dietary point of view”: Ibid.: 61.

“All those present”: Urbinati 1975.

Reconvened in Montreal: ISRF 1975 (“the impact of consumer advocates”: 6).

Recommendations of Errol Marliss: ISRF 1976.

“the effort to unite the world”: SAI 1977b.

“establish with the broadest possible”: SAI 1976.

Point one was the: Ibid.

“eminent and objective”: Tatem 1975.

“two strikingly polar attitudes”: Blackburn 1975.

“sugar critics”: Tatem 1976b.

Grande, Connor: Deutsch 1975.

Edwin Bierman: His role in shaping the ADA’s nutrition guidelines came about first via a paper on high-carbohydrate, low-fat diets for diabetics published in 1971, with John Brunzell as a collaborator (Brunzell et al. 1971), and then through his chairing of the ADA’s Committee on Food and Nutrition that same year, which was the first to begin liberalizing the recommended carbohydrate content of the diabetic diet (ADA 1971).

Involved in setting the diabetes research agenda: National Commission 1976: 81–105 ("argued eloquently": 96; "A review of all": 97).

"no known biological basis": Bierman 1979. Bierman's review chapter on carbohydrates and sugar was in a committee report of the American Journal of Clinical Nutrition, which was then used by administrators at the USDA to establish the first "Dietary Guidelines for Americans," released a year later.

Thirty research articles and reviews between 1952 and 1956: Cheek ed., 1974: 100–103.

"lead gift": Stare 1987: 175.

not even "remotely true": Whelan and Stare 1983: 194.

His department received funding: Stare 1987: 175–76.

Tobacco-industry documents reveal: See http://legacy.library.ucsf.edu/tid/qhn96b00/pdf for a description of the study, describing the conclusion before it was conducted — that body type could be blamed for heart disease rather than smoking. See http://legacy.library.ucsf

.edu/tid/eam96b00/pdf for Stare's request of funds for this study.

A martini at night: Hess 1978.

"and may be hazardous": Stare 1976a.

Sugar Association repeatedly turned to Stare: SAI 1975d.

"Sugar in the Diet of Man": Stare, ed., 1975.

Grande wrote the chapter: Grande 1975.

Bierman co-wrote: Bierman and Nelson 1975.

Twenty-five thousand copies: Darrow and Forrestal 1979: 739.

Included in their press packets . . . "falsely maligned": SAI 1975a: 2.

"Scientists Dispel Sugar Fears": SAI 1975b.

Funded entirely by the sugar industry . . . confidential memo: SAI 1975c.

"Professors on the Take": Rosenthal et al. 1976.

"A lot of the public": Hess 1978.

FDA would launch: On the history of the GRAS reviews, see USFDA 2015.

Seventy-two "comprehensive reports": LSRO 1977.

"Avoidance of even an appearance": Siu et al. 1977: 2530.

Irving . . . longtime member and chairman: ISRF 1969.

Fomen had received sugar-industry funding: Cheek, ed., 1974: 4.

"credible evidence . . . if sucrose was to be declared": Siu et al. 1977: 2534, 2535.

"urgent request . . . identify pertinent": Bollenbeck 1976.

"conflicting results": LSRO 1975: 7.

Cited fourteen such studies: These were references 30 and 46–58. Reference 56 was Grande's chapter; 46, 50, and 51 were from his laboratory; and 47 was funded by the sugar industry.

"suggest that long term consumption": This concerned reference 10.

Four contradictory reports: References 94–97: of those, 95 and 96 are studies from Bierman's laboratory, and reference 97 is his chapter with Nelson.

The revised version of the SCOGS review: LSRO 1976: 13–14.

"It is not possible": Ibid.: 14.

"contribut[ing] information": Ibid.: 29.

"proud of the credit": SAI 1977c: 2.

Reiser . . . and colleagues submitted: Ibid.: 30.

"abundant evidence": Reiser and Szepesi 1978.

"loudly proclaim[ing]": LSRO 1977: 2553.

"should be memorized": SAI 1977c: 2.

Footnote. "limitations of experimental design": Ibid.

"Sugar is Safe!": SAI 1977e.

Footnote. PRSA 1976.

Funding research on diabetes: SAI 1978: 13–43 ("prove of therapeutic value": 21).

"maintain research": SAI 1977d: 34.

Two researchers who received: Interviews, Ron Arky, Feb. 2, 2012; Paul Robertson, Jan. 6, 2012.

"would self destruct": SAI 1977a: 4.

"first comprehensive statement": Select Committee 1977.

"hammered away": SAI 1977a: 4.

"The weight given": McGovern 1977.

Hegsted later said: Interview, Mark Hegsted, March 30, 1999.

"Contrary to widespread opinion": USDA and HEW 1980.

Stated unambiguously: USDA and HEW 1985.

Come out of the USDA's own Carbohdyrate Nutrition Laboratory: Reiser et al. 1986 ("modest"); Reiser and Hallfrisch 1987.

"no conclusive evidence": Glinsmann et al. 1986: S15.

Surgeon General's Report: US HHS 1988 (linking sugar to chronic disease: 111).

Diet and Health: NRC 1989: 273–79.

Institute of Medicine: IOM 2005: 295–324.

"disproportionate consumption": Koop 1988.

Sugar Association . . . still misquoting: See http://www.sugar.org/sugar-your-diet/what-does-the-science-say/.

"when sugars are consumed": Glinsmann et al. 1986: S15.

Any substance could be harmful: Interview, Walter Glinsmann, Feb. 7, 2011.

Forty-two pounds of sugar per person: Glinsmann et al. 1986: S150–S216.

"played no causal role": COMA 1989: 43.

Chapter 9: What They Didn't Know

Epigraph. "I wish there were some formal courses": Thomas 1985.

"The method of science": Popper 1979: 81.

Hundred thousand subjects: Review Panel 1969, and US HEW 1971.

Quarter-billion dollars in two trials: MRFIT Research Group 1982; LRC Program 1984a; LRC Program 1984b.

"It's an imperfect world": Interview, Basil Rif kind, Aug. 6, 1999.

Massive public-relations campaign: See Taubes 2007: 58–61.

The authorities involved had little doubt: Marshall 1990.

Women's Health Initiative: Prentice et al.

2006 (breast cancer); Howard, Van Horn, et al. 2006 (heart disease and stroke); Howard, Manson, et al. 2006 (weight); Beresford et al. 2006 (colorectal cancer).

Chose *not* to perceive: See, for instance, NHLBI Communication Office 2006, Buzdar 2006, and WHO press release: http://www.who.int/nmh/media/Response_statement_16_feb_06F.pdf.

"the disproportionate consumption": Koop 1988.

The Cochrane Collaboration: Hooper et al. 2012.

"We're all being pushed": Interview, William Harlan, Jan. 24, 1999.

Footnote. Bacon 1994: 57.

Yudkin discussed this conflict: Yudkin 1971.

"strain specific": Bender and Damji 1971.

"just as great a mistake": Yudkin 1971.

A more nuanced perspective: On the biochemistry of sucrose and fructose, see, for instance, Shafrir 1991.

"unfettered by cellular controls": Lyssiotis and Cantley 2013.

"the most lipogenic": Interview, Walter Glinsmann, April 11, 2002.

"the remarkable hepatic": Shafrir 1991.

In human studies: See, for instance, Kraybill 1975, citing, among other studies, Roberts 1973.

Young women . . . relatively resistant: See, for instance, Nikkilä 1974.

Footnote. "shows a tendency": Higgins 1916.

Manifest . . . glucose intolerance: See, for instance, Bender and Damji 1971.

Cohen and his Israeli colleagues reported: Cohen et al. 1974.

Footnote. Interview, Walter Glinsmann, Feb. 7, 2011.

Researchers at Oxford University: Jenkins et al. 1981.

"for diabetics to be denied": Bantle et al. 1983.

Position of the American Diabetes Association: Vinik et al. 1987.

When 150 pounds of sugar sold: For sugar availability numbers, see the USDA Web site http://www.ers.usda.gov/data-products/food-availability-(per-capita)-data-system.aspx.

American Heart Association was recommending: Anon. 1995.

Referred to their product as "fructose": See, for instance, Anon. 1996: 16–18.

HFCS we were now consuming: For a good discussion of the role of HFCS in the food supply, see Duffey and Popkin 2008.

Footnote. "Invert sugar": Cantor 1975: 29.

Insulin resistance and . . . "metabolic

syndrome": See, for instance, Reaven 1988; Després et al. 1996; NHLBI 2015.

Seventy-five million adult Americans: Ervin 2009.

Reaven discussed the emerging science: Kolata 1987.

Reaven gave the prestigious Banting Lecture: Reaven 1988.

Large numbers of LDL particles: See, for instance, Hulthe et al. 2000.

Uric acid . . . chronic inflammation: See, for instance, Coutinho et al. 2007.

What causes the insulin resistance?: Taubes 2009.

"a marvelous model": Interview, Gerald Reaven, Dec. 9, 2010.

"ingested the contents": Zelman 1950.

First case reports: Ludwig et al. 1980 (in adults); Kinugasa et al. 1984 (in children).

One in every ten adolescents: Welsh et al. 2013.

Seventy-five million adults: NIDDK 2014b.

Established certain findings unambiguously: See, for instance, Tappy and Lê 2010.

Researchers say the metabolic effects: Interviews, Khosrow Adeli, Nov. 30, 2010; Luc Tappy, Dec. 2, 2010; Michael Paglisotti, Jan. 3, 2011; Claire Hollenbeck, Jan. 4, 2011; Peter Havel, Feb. 12, 2011.

"insulin resistance and many features":

Bremer et al. 2011.
"fascinated by the very peculiar metabolism": Interview, Luc Tappy, Dec. 2, 2010.
When the subjects lose weight: See, for instance, Rippe and Angelopoulos 2015.
Dedicated an entire issue: Nov. 1993.
"Further studies are clearly needed": Tappy and Jéquier 1993.
"clearly a need for intervention": Tappy and Lê 2010.
Fewer than a dozen clinical trials: From search on clinicaltrials.gov for "sucrose OR fructose AND United States."

Chapter 10: The If/Then Problem: I

Epigraph. "It is sometimes disheartening": Justice 1994.
Joslin traveled to Arizona: Joslin 1940.
One moment the Native American population seemed to be healthy: Justice 1994; interviews, David Pettitt, March 27, 2003; Peter Bennett, March 24, 2005; James Justice, April 7, 2005.
The Pima: For their history, see Russell 1975 ("The marvel is": 33); Smith et al. 1994 ("years of famine": 409): Taubes 2007: 235–39.
"largely bypassed": Price et al. 1993.

"critical juncture with modernity": Weidman 2012.

During the war years: Bernstein 1991 ("accelerated the detribalization process": 89).

Aleš Hrdlička commented: Hrdlička 1908: 156–57.

"exhibit a degree of obesity": Russell 1975: 66.

"everything obtainable": Hrdlička 1906.

"markedly flesh-producing": Russell 1975: 66.

Hrdlička had also weighed and measured: Hrdlička 1908: 347–48.

In 1938 . . . early 1940s . . . and 1949: Justice 1994.

Surveys done in the 1930s: Joslin 1940.

As late as 1947: Sugarman, Hickey, et al. 1990.

By the early 1950s: Kraus and Jones 1954 ("widespread poverty": 25; "That this obesity": 118).

Survey of inpatient records: Cohen 1954.

In 1954–55: Parks and Waskow 1961.

A disease they believed: Interview, Peter Bennett, March 24, 2005.

Over nine hundred Pima: Lawrence et al. 1966.

Reporting the results of the survey: Miller et al. 1965.

Bennett, Burch, and their colleagues were

confirming: Genuth et al. 1967; Bennett et al. 1971.
Studying the Papago and other local tribes: Justice 1994.
Clearly documented in the Navajo: Gohdes 1986.
Childhood obesity and type 2 diabetes: Sugarman, White, et al. 1990; Sugarman, Hickey, et al. 1990.
"shocked" by "the amount of suffering": Interview, Eric Ravussin, Feb. 22, 2005.
"As more thorough examinations": Justice 1994.
"fantastic opportunity": Interview, Peter Bennett, March 24, 2005.
Hrdlička had commented: Hrdlička 1906.
Similar to what rural Americans elsewhere: Darby et al. 1956.
"large amount of soft drinks": Hesse 1959.
USDA had initiated: Justice 1994.
"Even though evidence": Byers 1992.
"ration their children's sweets": Richardson 2002: 292–93.
Prior to the discovery of insulin: Feudtner 2003: 150.
prognosis for the mother "horrible": Joslin 1923: 649.
By the 1940s: Tattersall 2009: 94.
"they would then be fine": Interview, David Pettitt, March 27, 2003.

More than half of the children: Pettitt et al. 1983.

45 percent of the children: Pettitt et al. 1988.

"The baby is not diabetic": Interview, Boyd Metzger, Oct. 30, 2006.

Jorge Pedersen: On his hypothesis and its implications, see Catalano and Hauguel-De Mouzon 2010.

a "vicious cycle": Dabelea et al. 2000.

Alarming rise of diabetes internationally: Felita et al. 2006.

"general attitude of the medical profession": Allen 1913: 146.

Calls it a "myth": ADA 2015.

We can "save money": ADA 2014.

Accepts the role of fat accumulation: Geibel 2010.

"It is unknown": Pettitt et al. 1988.

Chapter 11: The If/Then Problem: II

Epigraph. Provisional List of Western Diseases: Trowell and Burkitt 1981: xv.

"one of the world's best-known": Auerbach 1974.

"It proved obnoxious": Trowell and Burkitt 1981: xvi.

"where the conditions of life": Chamberlain 1903.

"pattern and pathogenesis": Higginson 1997.

"Never before": Trowell 1981: 4.

Trowell and his colleagues experienced: Galton 1976 ("ancient Egyptians": 63).

Footnote. "Hundreds of x-rays": Galton 1976: 63.

First diagnosis of coronary heart disease: Trowell and Singh 1956.

"full of obese Africans": Trowell 1975.

"The incidence and variety of diseases": Trowell and Burkitt 1981: xiv.

"In relatively stable populations": Burkitt 1975.

"significance of relationships": Burkitt 1975.

"Before the spirochaete": Ibid.

"an extraordinary coincidence": Cleave 1975: 24.

"We are to admit": See https://en.wikiquote.org/wiki/Isaac_Newton.

The highest prevalence of diabetes: IDF 2013: 33; IDF 2015: 95. In the sixth edition of the IDF diabetes atlas, published in 2013, the prevalence of diabetes in Tokelau in adults (age twenty and older) is reported to be at 37.5 percent. In the seventh edition, the prevalence of "adult diabetes," apparently estimated for the whole population — those above and below the age of twenty — is given as 30 percent, still the world's highest.

More than two-thirds were obese: WHO Global Database on Body Mass Index, at http://apps.who.int/bmi/index.jsp.

Tokelau Island Migrant Study: Wessen et al., eds., 1992; Huntsman and Hooper 1996 (see pp. 1–20 for details of study; subsisted on a diet: 286–94); Wessen 2001.

Through the mid-1960s: Harding et al. 1986.

More than 50 percent of the calories: Prior et al. 1974.

Medical records of the islanders: Tuia 2001; Wessen et al., eds., 1992: 13.

A few had gout: Prior et al. 1987.

Women were diabetic: Østbye et al. 1989.

Change to a more Western dietary pattern: Wessen et al., eds., 1992: 288–89.

Changes for the Tokelauans who immigrated: Ibid.: 291–96; Harding et al. 1986.

Sugar consumption skyrocketed: Prior et al. 1978.

Diabetes prevalence shot: Østbye et al. 1989.

Gout also increased: Prior et al. 1987.

Obesity, unsurprisingly, also increased: Wessen et al., eds., 1992: 299.

Foods and drinks delivered: Rush and Pearce 2013.

Different dietary and lifestyle triggers: Wessen et al., eds., 1992: 383–88 ("different set of relevant variables": 384).

Egyptian mummies from: Newcombe 2013: 2.

Recent surveys suggest: See, for instance, Zhu et al. 2011.

Walking on one's eyeballs: Porter and Rousseau 1998: 3.

"a most regrettable circumstance": Bauer and Klemperer 1947.

a nearly vegetarian diet: Hydrick and Fox 1984.

"because of their ineffectiveness": Ibid.

incidence of gout in vegetarians: Bauer and Klemperer 1947 ("much higher than is generally assumed," and "largely vegetarians and teetotalers").

Eating *more* protein: Hydrick and Fox 1984.

In primitive populations: See, for instance, Benedek 1993; Trowell 1947.

Disease was so rare in East Africa: Benedek 1993; Beighton et al. 1977.

"one large gouty family": Rose 1975.

Higher rates of atherosclerosis and hypertension: Bauer and Klemperer 1947; Reaven 1997.

Diabetes is also commonly associated with gout: See, for instance, Buchanan 1972; Whitehouse and Cleary 1966.

In 1951, Harvard researchers: Gertler et al. 1951.

Investigators first linked hyperuricemia: Reiser 1987; Reaven 1997.

"gout wave": Wyngaarden and Kelley, eds., 1976: ix.

"a luxury of kings": Mintz 1985: 96. For a good history of gout and how it spread, see Porter and Rousseau 1998.

Finnish researchers, who referred: Perheentupa and Raivio 1967.

When fructose is metabolized in the liver: See, for instance, Mayes 1993; Hydrick and Fox 1984.

"a fairly common cause of gout": Seegmiller et al. 1990.

"Since serum-uric-acid levels": Perheentupa and Raivio 1967.

"fructose can accelerate": Hydrick and Fox 1984.

High-fructose diets in healthy individuals: Mayes 1993.

The major players had left the field: Interviews, Irving Fox, May 18, 2004; Peter Mayes, May 26, 2004; Thomas Benedek, June 14, 2004; James Seegmiller, August 5, 2004; William Kelley, Aug. 6, 2004.

"sugars" and "sweets" as among the recommended foods: See, for instance, Fam 2002; Emmerson 1996.

Richard Johnson, a kidney specialist: See, for instance, Johnson et al. 2007; Feig et al. 2008.

Hypertension is yet another example: Kotchen 2011.

"a saving process": Warfield 1920: 106.

"claims would have to be paid": Symonds 1923.

After another twenty years: For reviews of the early literature on hypertension and isolated populations, see Kean and Hammill 1949; Lowenstein 1954.

In the Philippines: Discussed in Shattuck 1937.

Among Zuni Indians: Fleming 1924.

Inuit in Greenland and Labrador: Thomas 1928.

Native tribes in Kenya: Donnison 1929 ("This contrast").

Bedouin tribes in Syria: Hudson and Young 1931 ("the conspicuous hypotension").

The Yucatán and Guatemala: Shattuck 1937.

Among Kuna Indians: Kean 1944 ("a striking finding").

By the 1960s: Trowell 1981.

Two tribes of Brazil Indians: Lowenstein 1961.

In fifty-two communities: Intersalt 1988.

Salt was not just: See also Page et al. 1974.

Footnote. Kenya and Uganda: Shaper 1967; Shaper et al. 1969. South Pacific Islanders: Prior et al. 1964; Prior 1971.

"As soon as we think": Schulz 2010: 310.

Salt/hypertension hypothesis: For systematic reviews of the evidence, see He et al. 2013; Graudal et al. 2011.

"deadly white powder": Jacobson 1978.

Carl von Voit suggested: In Rony 1940: 154.

Confirmed this observation: Benedict et al. 1919: 195.

By 1933: Atchley et al. 1933.

Insulin was being implicated: A good review is DeFronzo 1981 ("insulin, working through sodium").

"Antidiuresis associated with": Miller and Bogdonoff 1954.

Landsberg . . . discovered: Landsberg 1986; Landsberg 2001.

Richard Johnson's work: Johnson et al. 2007.

Salt sensitivity is an elusive: See, for instance, Lastra et al. 2010; Luzardo et al. 2015.

Caused in rats: Johnson et al. 2002.

Salt sensitivity is caused by insulin resistance: Yatabe et al. 2010; Laffer and Elijovich 2013.

"like insanity": Tanchou 1844: 263.

Cancer Research Fund: Dukes 1964.

"all matters connected": Anon. 1902.

"placed in formalin": Elgin 1906.

Letters and specimens began to arrive: See, for instance, Anon. 1906.

"There is a general unanimity": Moffat 1904.

The fund's . . . published its third report: Bashford 1908a.

"almost universal endeavor": Bashford 1908b: 9.

"wholesome tendency": Fitz and Joslin 1898.

"serve no useful purpose": Bashford 1908b.

1910 and then again in 1915: Levin 1910; Hoffman 1915: 151.

Half a century later: Thomas 1979; Sorem 1985; Bleed et al. 1992; interview, James Justice, April 7, 2005.

Hoffman published his: Hoffman 1915 ("qualified medical observers": 147).

"There are no known reasons": Ibid.

"one of the few diseases": Ibid.: 4.

"By the 1930s": WCRF and AICR 1997: 36.

"astonished to encounter . . . the natives": Schweitzer 1957.

John Higginson: His studies are reviewed in Higginson 1981 and Higginson 1997.

"potentially preventable": Doll and Peto 1981.

Single-case reports in medical journals: Brown et al. 1952.

Canadian physicians published an analysis: Hildes and Schaefer 1984.

"only a very small part": Higginson 1983.

Japanese women: See, for instance, Buell 1973; Ziegler et al. 1993.

Footnote. Marmot and Syme 1976.

Seminal article on: Doll and Peto 1981.

"the coincidence of diabetes": Anon. 1889.

From the Centers for Disease Control: Calle et al. 2003.

Linking cancer to diabetes: Coughlin et al. 2004.

Cancer seems to thrive: see Taubes 2012.

Link between cancer and insulin: Giovannucci 1995; Kaaks 1996; Burroughs et al. 1999; Kaaks and Lukanova 2001; LeRoith and Roberts 2003; Pollak et al. 2004. More recent reviews include Taubes 2012; Poloz and Stambolic 2015.

Scottish researchers reported: Evans et al. 2005.

Association confirmed multiple times: Noto et al. 2012.

Researchers — including Howard Temin: Temin 1967; Temin 1968.

"addicted to" insulin: Taubes 2012.

"intensely stimulated cell": Heusen et al. 1967.

"exquisitely sensitive to insulin": Osborne et al. 1976.

As much a metabolic as "proliferative": See, for instance, Coller 2014; Bowers et al. 2015.

The Warburg effect: Vander Heiden et al. 2009.

Free radicals . . . mutate DNA: Interview, Craig Thompson, Feb. 1, 2011.

As Cantley has said: Interview, Lewis Cantley, Feb. 1, 2011.

Alzheimer's disease was only officially recognized: Ingram 2015: 24–29.

Residents of Hisayama, Japan: Yoshitake et al. 1995.

Rotterdam, the Netherlands: Ott et al. 1996.

Rochester, Minnesota: Leibson et al. 1997.

"direct or indirect": Ott et al. 1999.

Several studies have shown: See, for instance, Li et al. 2015.

High blood sugar . . . AGEs: See, for instance, Umegaki 2014.

Alzheimer's as type 3 diabetes: See, for instance, Guthrie 2007.

"is vital in the fine-tuning": Kleinridders et al. 2014.

Seminal study of nuns: Snowdon et al. 1997. For more recent confirmation of these results, see, for instance, Vermeer et al. 2003; Schneider et al. 2007.

"the direction of the relationship": Castro et al. 2014.
"it remains to be determined": Barlow et al. 2015.
Since the 1950s: Ahrens 1957.
"It would be an extraordinary coincidence": Cleave 1975: 24.
We should "eat food": Pollan 2008: 1.

Epilogue: How Little Is Still Too Much?

"No child can grow up": Feudtner 2003: 133.
"people just act like it is": Mann 2011: 289.
"This definite incrimination": Allen 1913: 147.
"inclining to be too fat": Slare 1915: E4.
"more quickly and surely": Brillat-Savarin 1986: 240.
"skeleton-like appearance": Brigham 1868.
"foodlike substances": Pollan 2008: 1.
Artificial sweeteners . . . metabolic syndrome: See, for instance, Bruyère et al. 2015.
Footnote. Diet known as DASH: Appel et al. 1997.
Sweet-taste receptors in our guts: Fernstrom et al. 2012.

BIBLIOGRAPHY

Abbott, E. 2007. *Sugar: A Bittersweet History.* Toronto: Penguin Canada.

Abel, M. H. 1915. *Sugar and Its Value as Food.* Farmers Bulletin 535, U.S. Department of Agriculture. Washington, D.C.: Government Printing Office.

Abraham, C. 2011. "How the Diabetes-Linked 'Thrifty Gene' Triumphed with Prejudice Over Proof." *Globe and Mail,* Feb. 25. At http://www.theglobeandmail.com/news/national/how-the-diabetes-linked-thrifty-gene-triumphed-with-prejudice-over-proof/article569423/?page=all.

Ahmed, S. H. 2012. "Is Sugar as Addictive as Cocaine?" In *Food and Addiction: A Comprehensive Handbook,* ed. K. D. Brownell and M. S. Gold (Oxford, U.K.: Oxford University Press), 231–38.

Ahrens, E. H., Jr. 1957. "Nutritional Fac-

tors and Serum Lipid Levels." *American Journal of Medicine* 23, no. 6 (Dec.): 928–52.
Ahrens, E. H., Jr., J. Hirsch, W. Insull, Jr., T. T. Tsaltas, R. Blomstrand, and M. L. Peterson. 1957. "Dietary Control of Serum Lipids in Relation to Atherosclerosis." *J.A.M.A.* 164, no. 17 (Aug. 24): 1905–11.
Ahrens, E. H., Jr., J. Hirsch, K. Oette, J. W. Farquhar, and Y. Stein. 1961. "Carbohydrate-Induced and Fat-Induced Lipemia." *Transactions of the Medical Society of London* 74: 134–46.
Ahrens, E. H., Jr., W. Insull, Jr., R. Blomstrand, J. Hirsch, T. T. Tsaltas, and M. L. Peterson. 1957. "The Influence of Dietary Fats on Serum-Lipid Levels in Man." *Lancet* 272 (May 11): 943–53.
Albrink, M. J. 1965. "Diet and Cardiovascular Disease." *Journal of the American Dietetic Association* 46 (Jan.): 26–29.
———. 1963. "The Significance of Serum Triglycerides." *Journal of the American Dietetic Association* 42 (Jan.): 29–31.
Albrink, M. J., P. H. Lavietes, E. B. Man, and J. R. Paul. 1962. "Relationship Between Serum Lipids and the Vascular Complications of Diabetes from 1931 to 1961." *Transactions of the Association of*

American Physicians 75: 235–41.
Alcoholics Anonymous (AA). 2001. *Alcoholics Anonymous,* 4th edition. Alcoholics Anonymous. At http://2travel.org/Files/AA/BigBook.pdf.
Allen, F. M. 1913. *Studies Concerning Glycosuria and Diabetes.* Cambridge, Mass.: Harvard University Press.
Alonso, L. G., and T. H. Maren. 1955. "Effect of Food Restriction on Body Composition of Hereditary Obese Mice." *American Journal of Physiology* 183, no. 2 (Oct.): 284–90.
American Cancer Society (ACS). 2016. "Lifetime Risk of Developing and Dying from Cancer." At http://www.cancer.org/cancer/cancerbasics/lifetime-probability-of-developing-or-dying-from-cancer.
American Diabetes Association (ADA). 2016. "Statistics about Diabetes." At http://www.diabetes.org/diabetes-basics/statistics/.
———. 2015. "Diabetes Myths." At http://www.diabetes.org/diabetes-basics/myths/.
———. 2014. "Healthy Eating." At http://www.diabetes.org/are-you-at-risk/lower-your-risk/healthy-eating.html.
———. 2013. "Economic Costs of Diabetes in the U.S. in 2012." *Diabetes Care* 36, no. 4 (March 14): 1033–46.

———. 1971. "Principles of Nutrition and Dietary Recommendations for Patients with Diabetes Mellitus: 1971." *Diabetes* 20, no. 9 (Sept.): 633–34.

American Heart Association (AHA). 1961. "Dietary Fat and Its Relation to Heart Attacks and Strokes: Report by the Central Committee for Medical and Community Program of the American Heart Association." *J.A.M.A.* 175, no. 5 (Feb. 4): 389–91.

American Lung Association (ALA), Epidemiology and Statistics Unit. 2014. "Trends in Lung Cancer Morbidity and Mortality." November. At http://www.lung.org/assets/documents/research/lc-trend-report.pdf.

Anderson, J. T., F. Grande, Y. Matsumoto, and A. Keys. 1963. "Glucose, Sucrose and Lactose in the Diet and Blood Lipids in Man." *Journal of Nutrition* 79 (March): 349–59.

Anitschkow, N., and S. Chalatow. 1913. "Über experimentelle Cholesterinsteatose und ihre Bedeutung für die Entstehung einiger pathologischer Prozesse." *Centrbl Allg Pathol Pathol Anat.* 24 (1913): 1–9.

Anon. 2016. "Bottled and Canned Soft Drinks and Carbonated Water." Highbeam Business. At https://business.highbeam

.com/industry-reports/food/bottled-canned-soft-drinks-carbonated-waters.
———. 1996. *Corn Annual.* Corn Refiners Association, Inc.
———. 1995. *An Eating Plan for Healthy Americans: The American Heart Association Diet.* Dallas: American Heart Association.
———. 1989. "AHA Conference Report on Cholesterol." *Circulation* 80, no. 3 (Sept.): 715–48.
———. 1979. "Julius Bauer." *Lancet* 313 (June 23): 1359.
———. 1964. "Merchandising: Bubbling Along." *Time,* Aug. 7. At http://content.time.com/time/subscriber/article/0,33009,871356,00.html.
———. 1961. "The Fat of the Land." *Time,* Jan. 13: 48–52.
———. 1956. "News of the Advertising and Marketing Fields." *New York Times,* July 26: 32.
———. 1955a. "Calculating Calories." *Forbes,* Oct.: 22.
———. 1955b. "Sugar Bowled Over by Photo." *New York Times,* Aug. 15: 4.
———. 1955c. "Combined Staff Clinic: Obesity." *American Journal of Medicine* 19, no. 1 (July): 111–25.
———. 1954. "News of the Advertising and

Marketing Fields." *New York Times,* Jan. 12: 38.

———. 1953. "Modern Living: Battle of the Bulge." *Time,* Aug. 10. At http://content.time.com/time/magazine/article/0,9171,818679,00.html.

———. 1951a. "Little Known Sugar Facts." *New York Amsterdam News,* Sept. 29: 21.

———. 1951b. "To Stress Sugar for Energy." *New York Times,* April 28: 31.

———. 1948a. "Reports of Local Heart Association Activities." *American Heart Journal* 36: 158–59.

———. 1948b. "National Heart Week." *American Heart Journal* 35: 528.

———. 1945a. "The Bitter End." *Time,* Oct. 8. At http://content.time.com/time/subscriber/article/0,33009,776288,00.html.

———. 1945b. "Additional Grants of the Sugar Research Foundation." *Science* 101 (Feb. 2): 110–11.

———. 1944a. "War Seen Changing Our Eating Habits." *New York Times,* Oct. 4: 22.

———. 1944b. "100,000,000 Pounds of Candy for Army." *New York Times,* June 7: 22.

———. 1943. "The Sugar Research Foundation." *Science* 98, no. 2,554 (Dec. 10):

509–10.
———. 1942a. “Sugar Rationing Called a ‘Godsend’ to National Health.” *Science News Letter* 41, no. 11 (March 14): 164.
———. 1942b. “Scientists Are Offered $45,000 to Find New Uses for Sugar.” *Boston Globe,* March 3: 2.
———. 1939. “Professor of Medicine Augments Teaching with Research.” *Michigan Alumnus* 45 (June 10): 415.
———. 1936a. “Sugar Institute Closes; Main Activities Banned.” *New York Times,* Nov. 19: 39.
———. 1936b. “Find Trust Abuses in Sugar Institute.” *New York Times,* March 30: 1.
———. 1934. “Advises Reducing Sugar in Diet to Avoid Tooth Decay.” *Science News Letter* 26 (Nov. 10): 300.
———. 1932. “Starts Suit to End Sugar Institute.” *New York Times,* Feb. 10; 33.
———. 1931. “Business: Chadbourne Home.” *Time,* Feb. 2. At http://content.time.com/time/ subscriber/article/0,33009,740959-1,00.html.
———. 1929. “Trim Figure Mode, Sugar Crisis Factor.” *New York Times,* April 5: 6.
———. 1928a. “Americans Saturated with Sugar.” *Science News Letter,* Dec. 22: 329.
———. 1928b. “Sugar Institute Is Orga-

nized Here." *New York Times*, Jan. 8:43.
———. 1926. "Use of Sugar by Crews Not New, Says Stevens." *New York Times*, March 30: 28.
———. 1925a. "Sugar and Athletics." *Lancet* 206 (Sept. 19): 611.
———. 1925b. "Tells of Big Drop in Our Use of Whisky." *New York Times*, July 18: 5.
———. 1925c. "Blames Auto for Diabetes Spread." *Boston Globe*, May 13: 23.
———. 1925d. "Sees Champions Made by Chocolate Bars." *New York Times*, March 16: 19.
———. 1924. "Yale Soccer Team Eats Sugar to Increase Energy, but Loses." *New York Times*, Nov. 11: 28.
———. 1923. "War on Diabetes." *Time*, April 21: 20.
———. 1921a. "Columbus Brought First Sugar Cane." *New York Times*, June 26: 21.
———. 1921b. "To Be Record Year in Use of Sugar." *New York Times*, June 19: 24.
———. 1920. "Candy Stores Get Old Saloon Trade." *New York Times*, Feb. 22: 23.
———. 1919a. "Scarcity in Sugar Puzzles Officials." *New York Times*, Oct. 19: 46.
———. 1919b. "Much Food Value in Soft Drinks." *New York Times*, May 25: 27.

———. 1910. "Calls Sugar a Human Bane." *New York Times,* July 22: 1.

———. 1909. "Concerning Sugar as a Cure for Inebriety." *New York Times,* Feb. 28: 51.

———. 1906. Papers Relating to Cancer Research. In Parliamentary Papers: 1850–1908. Volume 53. Great Britain: Parliament. House of Commons.

———. 1903. "Candy Trade's Growth." *New York Times,* Dec. 20: 18.

———. 1902. "The Royal Colleges and the Investigation of Cancer." *Lancet* 159 (April 19): 1131–32.

———. 1889. "Diabetes and Tumours." *British Medical Journal* 1, no. 1,468 (Feb. 16): 376.

———. 1887. "Saccharin." *British Medical Journal* 2, no. 1,398 (Oct. 15): 838–39.

———. 1886. Editorial article 6. *New York Times,* Sept. 17: 4.

———. 1884. "Suppose We Had No Sugar." *New York Times,* Dec. 21: 11.

———. 1873. "House of Commons." *Pall Mall Budget,* April 10: 28.

———. 1857. "Discouraging for Sugar Consumers." *New York Times,* May 22:4.

———. 1856. "Sugar." *New York Times,* Nov. 14: 4.

Appel, L. J., T. J. Moore, E. Obarzanek, et al. 1997. "A Clinical Trial of the Effects of Dietary Patterns on Blood Pressure." *New England Journal of Medicine* 336, no. 16 (April 17): 1117–24.

Aretaeus of Cappadocia. 1837. "On Diabetes." Trans. T. F. Reynolds. In *Diabetes: A Medical Odyssey* (Tuckahoe, N.Y.: USV Pharmaceutical Corp., 1971), 1–6.

Atchley, D. W., R. F. Loeb, D. W. Richards, Jr., E. M. Benedict, and M. E. Driscoll. 1933. "On Diabetic Acidosis: A Detailed Study of Electrolyte Balances Following the Withdrawal and Reestablishment of Insulin Therapy." *Journal of Clinical Investigation* 12, no. 2 (March 1): 297–326.

Atwater, W. O. 1888. "What We Should Eat." *Century Illustrated Magazine.* 36 no. 2 (June): 257.

Auerbach, S. 1974. "Roughing It — Tonic for Our Time." *Washington Post,* Aug 19: B1.

Avena, N. M, P. Rada, and B. G. Hoebel. 2008. "Evidence for Sugar Addiction: Behavioral and Neurochemical Effects of Intermittent, Excessive Sugar Intake." *Neuroscience and Biobehavioral Reviews* 32, no. 1 (May 18): 20–39.

Aykroyd, W. R. 1967. *The Story of Sugar.* Chicago: Quadrangle Books.

Babst, E. D. 1940. *Occasions in Sugar.* New York: privately printed.

Bacon, F. 1994. *Novum Organum.* Ed. and trans. P. Urbach and J. Gibson. Peru, Ill.: Carus Publishing Company. [Originally published in 1620.]

Bahner, F. 1955. "Fettsucht und Magersucht." In F. Bahner, H. W. Bansi, G. Fanconi, A. Jores, and W. Zimmerman, eds. *Innersekretorische Krankheiten Fettsucht Magersucht,* ed. F. Bahner, H. W. Bansi, G. Fanconi, A. Jores, and W. Zimmerman. Vol. VII, no. 1 of *Handbuch der Inneren Medizin,* 4th edition (Berlin: Springer-Verlag), 978–1163.

Bantle, J. P., D. C. Laine, G. W. Castle, J. W. Thomas, B. J. Hoogwerf, and F. C. Goetz. 1983. "Postprandial Glucose and Insulin Responses to Meals Containing Different Carbohydrates in Normal and Diabetic Subjects." *New England Journal of Medicine* 309, no. 1 (July 7): 7–12.

Barker, T. C., D. J. Oddy, and J. Yudkin. 1970. *The Dietary Surveys of Dr Edward Smith 1862–3: A New Assessment.* London: Staples Press.

Barlow, G. M., A. Yu, and R. Mathur. 2015. "Role of the Gut Microbiome in Obesity and Diabetes Mellitus." *Nutrition in Clinical*

Practice 30, no. 6 (Dec.): 787–97.
Barnard, E. F. 1928. "Too Much Sugar for the World to Eat." *New York Times,* April 8: 112–14.
Bart, P. 1962. "Advertising: Calorie Craze and Its Impact." *New York Times,* Feb. 25: F12.
Bashford, E. F. 1908a. *Third Scientific Report on the Investigations of the Imperial Cancer Research Fund.* London: Taylor and Francis.
———. 1908b. "The Ethnological Distribution of Cancer." In Bashford 1908a, 1–26.
Bauer, J. 1941. "Obesity: Its Pathogenesis, Etiology and Treatment." *Archives of Internal Medicine* 67, no. 5 (May): 968–94.
———. 1940. "Some Conclusions from Observations on Obese Children." *Archives of Pediatrics* 57: 631–40.
———. 1929. "Endogene Fettsucht." *Verhandl. d. deutsch. Gesellsch. f. Verdauungs-u. Stoffechselkr* 9: 116. Cited in Bauer 1941.
Bauer, W., and F. Klemperer. 1947. "Gout." In *Diseases of Metabolism,* ed. G. G. Duncan (Philadelphia: W. B. Saunders), 609–56.
Beighton, P., L. Solomon, C. L. Soskolne, and M. B. E. Sweet. 1977. "Rheumatic

Disorders in the South African Negro: Part IV, Gout and Hyperuricaemia." *South African Medical Journal,* June 25, 1969–72.

Belair, F., Jr. 1937. "Sugar Again Causes Legislative Battle." *New York Times,* Aug. 8: 7.

Bender, A. E., and K. B. Damji. "Some Effects of Dietary Sucrose." 1971. In Yudkin, Edelman, and Hough, eds., 1971, 172–82.

Bender, A. E., K. B. Damji, M. A. Khan, I. H. Khan, L. McGregor, and J. Yudkin. 1972. "Sucrose Induction of Hepatic Hyperplasia in the Rat." *Nature* 238 (Aug. 25): 461–62.

Benedek, T. G. 1993. "Gout." In *The Cambridge World History of Human Disease,* ed. K. F. Kiple (Cambridge, U.K.: Cambridge University Press), 763–72.

Benedict, F. G., W. R. Miles, P. Roth, and H. M. Smith. 1919. *Human Vitality and Efficiency Under Prolonged Restricted Diet.* Washington, D.C.: Carnegie Institution of Washington.

Bennett, P. H., T. A. Burch, and M. Miller. 1971. "Diabetes Mellitus in American (Pima) Indians." *Lancet* 298 (July 17): 125–28.

Beresford, S. A., K. C. Johnson, C. Ritenbaugh, et al. 2006. "Low-Fat Dietary Pattern and Risk of Colorectal Cancer: The Women's Health Initiative Randomized Controlled Dietary Modification Trial." *J.A.M.A.* 295, no. 6 (Feb. 8): 643–54.

Bergman, G. von, and F. Stroebe. 1927. "Die Fettsucht." In *Handbuch der Biochemie des Menschen und der Tiere,* ed. C. Oppenheimer (Jena, Germany: Verlag von Gustav Fischer), 562–98.

Bernstein, A. R. 1991. *American Indians and World War II.* Norman: University of Oklahoma Press.

Bernstein, L. M., and M. I. Grossman. 1956. "An Experimental Test of the Glucostatic Theory of Regulation of Food Intake." *Journal of Clinical Investigation* 35, no. 6 (June): 627–33.

Berson, S. A., and R. S. Yalow. 1965. "Some Current Controversies in Diabetes Research." *Diabetes* 14, no. 9 (Sept.): 549–72.

Bierman, E. L. 1979. "Carbohydrate and Sucrose Intake in the Causation of Atherosclerotic Heart Disease, Diabetes Mellitus, and Dental Caries." Supplement, *American Journal of Clinical Nutrition* 32, no. 12 (Dec.): 2644–47.

Bierman, E. L., and R. Nelson. 1975. "Carbohydrates, Diabetes, and Blood Lipids." *World Review of Nutrition and Di-*

etetics 22: 280–87.
Blackburn, H. n.d. "Ancel Keys." At http://mbbnet.umn.edu/firsts/blackburn_h.html.
———. 1975. "Contrasting Professional Views on Atherosclerosis and Coronary Disease." *New England Journal of Medicine* 292, no. 2 (Jan. 9): 105–7.
Blass, E. M. 1987. "Opioids, Sweets and a Mechanism for Positive Affect: Broad Motivational Implications." In *Sweetness,* ed. J. Dobbing (Berlin: Springer-Verlag), 115–24.
Bleed, D. M., D. R. Risser, S. Sperry, D. Hellhake, and S. D. Helgerson. 1992. "Cancer Incidence and Survival Among American Indians Registered for Indian Health Service Care in Montana, 1982–1987." *Journal of the National Cancer Institute* 84, no. 19 (Oct. 7): 1500–1505.
Bluher, M., B. B. Kahn, and C. R. Kahn. 2003. "Extended Longevity in Mice Lacking the Insulin Receptor in Adipose Tissue." *Science* 288 (Jan. 24): 572–74.
Bollenbeck, G. N. 1976. "Letter to Heads of Member Companies, Public Communications Committee. Subject: Tentative Evaluation of the Health Aspects of Sucrose as a Food Ingredient." Washington, D.C., Jan. 30. Sugar Association, Inc., Records of the Great Western Sugar Com-

pany. Colorado Agricultural Archive, Colorado State University.
Borders, W. 1965. "New Diet Decried by Nutritionists." *New York Times,* July 7: 16.
Borrell, B., and R. C. Duncan. 1993. "A Survey of World Sugar Policies." In Marks and Maskus, eds., 1993, 15–48.
Bowers, L. W., E. L. Rossi, C. H. O'Flanagan, L. A. de Graffenreid, and S. D. Hursting. 2015. "The Role of the Insulin/IGF System in Cancer: Lessons Learned from Clinical Trials and the Energy Balance–Cancer Link." *Frontiers in Endocrinology* 6 (May): 1–16.
Bramen, L. 2010. "The Evolution of the Sweet Tooth." *Smithsonian.com.* Feb. 10. At http://www.smithsonianmag.com/ arts-culture/the-evolution-of-the-sweet-tooth-79895734/?no-ist.
Braudel, F. 1992. *Civilization and Capitalism, 15th–18th Century: The Wheels of Commerce.* Berkeley: University of California Press.
Bray, G. A., S. J. Nielsen, and B. M. Popkin. 2004. "Consumption of High-Fructose Corn Syrup in Beverages May Play a Role in the Epidemic of Obesity." *American Journal of Clinical Nutrition* 79, no. 4 (April): 537–43.
Bremer, A. A., K. L. Stanhope, J. L. Gra-

ham, et al. 2011. "Fructose-Fed Rhesus Monkeys: A Nonhuman Primate Model of Insulin Resistance, Metabolic Syndrome, and Type 2 Diabetes." *Clinical and Translational Science* 4, no. 4 (August): 243–52.

Brigham, C. B. 1868. "An Essay upon Diabetes Mellitus." In *Diabetes: A Medical Odyssey* (Tuckahoe, N.Y.: USV Pharmaceutical Corp., 1971), 71–107.

Brillat-Savarin, J. A. 1986. *The Physiology of Taste.* Trans. M. F. Fisher. San Francisco: North Point Press. [Originally published 1825.]

Brody, J. E. 1977. "Sugar: Villain in Disguise?" *New York Times,* May 25: C1.

Brooks, C. M. 1946. "The Relative Importance of Changes in Activity in the Development of Experimentally Produced Obesity in the Rat." *American Journal of Physiology* 147, no. 4 (Dec.): 708–16.

Brooks, C. M., and E. F. Lambert. 1946. "A Study of the Effect of Limitation of Food Intake and the Method of Feeding on the Rate of Weight Gain During Hypothalamic Obesity in the Albino Rat." *American Journal of Physiology* 147, no. 4 (Dec.): 695–707.

Brown, G. M., L. B. Cronk, and T. J. Boag.

1952. "The Occurrence of Cancer in an Eskimo." *Cancer* 5, no. 1 (Jan.): 142–43.

Bruce, S. and B. Crawford. 1995. *Cerealizing America: The Unsweetened Story of American Breakfast Cereal.* Winchester, Mass.: Faber and Faber.

Brunzell, J. D., R. L. Lerner, W. R. Hazzard, D. Porte, Jr., and E. L. Bierman. 1971. "Improved Glucose Tolerance with High Carbohydrate Feeding in Mild Diabetes." *New England Journal of Medicine* 284, no. 10 (March 11): 521–24.

Bruyère, O., S. H. Ahmed, C. Atlan, et al. 2015. "Review of the Nutritional Benefits and Risks Related to Intense Sweeteners." *Archives of Public Health* 73 (Oct. 1): 41.

Buchanan, K. D. 1972. "Diabetes Mellitus and Gout." *Seminars in Arthritis and Rheumatism* 2, no. 2 (Fall): 157–62.

Buell, P. 1973. "Changing Incidence of Breast Cancer in Japanese-American Women." *Journal of the National Cancer Institute* 51, no. 5 (Nov.): 1479–83.

Burkitt, D.P. 1975. "Significance of Relationships." In Burkitt and Trowell, eds., 1975, 9–20.

Burkitt, D. P., and H. C. Trowell, eds. 1975. *Refined Carbohydrate Foods and Disease: Some Implications of Dietary Fibre.* New York: Academic Press.

Burroughs, K. D., S. E. Dunn, J. C. Barrett, and J. A. Taylor. 1999. "Insulin-Like Growth Factor I: A Key Regulator of Human Cancer Risk?" *Journal of the National Cancer Institute* 91, no. 7 (April 7): 579–81.

Burrows, E. G., and M. Wallace. 1999. *Gotham: A History of New York City to 1898.* New York: Oxford University Press.

Buzdar, A. U. 2006. "Dietary Modification and Risk of Breast Cancer." *J.A.M.A.* 295, no. 6 (Feb. 8): 691–92.

Byers, T. 1992. "The Epidemic of Obesity in American Indians." *American Journal of Diseases of Children* 146, no. 3 (March): 285–86.

Cahill, G. F., Jr. 1978. "Obesity and Diabetes." In *Recent Advances in Obesity Research:* vol. II, ed. G. A. Bray (London: Newman Publishing, 1978), 101–10.

Calle, E. E., C. Rodriguez, K. Walker-Thurmond, and M. J. Thun. 2003. "Overweight, Obesity, and Mortality from Cancer in a Prospectively Studied Cohort of U.S. Adults." *New England Journal of Medicine* 348, no. 17 (April 24): 1625–38.

Campbell, G. D. 1963. "Diabetes in Asians and Africans in and Around Durban." *South African Medical Journal* 37 (Nov.

30): 1195–1208.
Cantor, S. M. 1975. "Patterns of Use." In NAS 1975, 19–35.
Castro, A. V., C. M. Kolka, S. P. Kim, and R. N. Bergman. 2014. "Obesity, Insulin Resistance and Comorbidities — Mechanisms of Association." *Arquivos Brasileiros de Endocrinologia e Metabologia* 58, no. 6 (Aug.): 600–609.
Catalano, P. M., and S. Hauguel–De Mouzon. 2010. "Is It Time to Revisit the Pedersen Hypothesis in the Face of the Obesity Epidemic?" *American Journal of Obstetrics and Gynecology* 204, no. 6 (June): 479–87.
Centers for Disease Control and Prevention (CDC). 2016a. "Chronic Disease Overview." At http://www.cdc.gov/chronicdisease/overview/.
———. 2016b. "Obesity and Overweight." At http://www.cdc.gov/nchs/fastats/obesity-overweight.htm.
———. 2014a. "Long-Term Trends in Diabetes." October. At http://www.cdc.gov/diabetes/statistics.
———. 2014b. "National Diabetes Statistics Report, 2014." At http://www.cdc.gov/diabetes/pubs/statsreport14/national-diabetes-report-web.pdf.
Chamberlain, J. 1903. Mr. Chamberlain to

Governors and High Commissioners of Crown Colonies and Protectorates. In *Correspondence Relating to Cancer Research* (London: His Majesty's Stationery Office, 1905), 6.

Charles, R. H. 1907. "Discussion on Diabetes in the Tropics." *British Medical Journal* 2 (Oct. 19): 1051–64.

Chaudhuri, S., and M. Esterl. 2016. "U.K. Unveils Levy on Sugary Drinks." *The Wall Street Journal,* March 16. At http://www.wsj.com/articles/u-k-unveils-levy-on-sugary-drinks-1458144731.

Cheek, D. W,. ed. 1974. *Sugar Research 1943–1972.* International Sugar Research Foundation, Inc.

Cleave, T. L. 1975. *The Saccharine Disease: The Master Disease of Our Time.* New Canaan, Conn.: Keats Publishing.

———. 1956. "The Neglect of Natural Principles in Current Medical Practice." *Journal of the Royal Naval Medical Service* 42, no. 2 (Spring): 55–82.

———. 1940. "Instincts and Diet." *Lancet* 235 (April 27): 809.

Cleave, T. L., and G. D. Campbell. 1966. *Diabetes, Coronary Thrombosis and the Saccharine Disease.* Bristol, U.K.: John Wright & Sons.

Cohen, A. M. 1963. "Effect of Environmental Changes on Prevalence of Diabetes and of Atherosclerosis in Various Ethnic Groups in Israel." In *The Genetics of Migrant and Isolate Populations*, ed. E. Goldschmidt (New York: Williams & Wilkins), 127–30.

Cohen, A. M., S. Bavly, and R. Poznanski. 1961. "Change of Diet of Yemenite Jews in Relation to Diabetes and Ischaemic Heart-Disease." *Lancet* 278 (Dec. 23): 1399–1401.

Cohen, A. M., A. Teitelbaum, S. Briller, L. Yanko, E. Rosenmann, and E. Shafrir. 1974. "Experimental Models of Diabetes." In Sipple and McNutt, eds., 1974, 484–511.

Cohen, B. M. 1954. "Diabetes Mellitus Among Indians of the American Southwest: Its Prevalence and Clinical Characteristics in a Hospitalized Population." *Annals of Internal Medicine* 40, no. 3 (March): 588–99.

Cohen, P., M. Miyazaki, N. D. Socci, et al. 2002. "Role for Stearoyl-CoA Desaturase-1 in Leptin-Mediated Weight Loss." *Science* 297 (July 12): 240–43.

Cohen, R. 2013. "Sugar Love." *National Geographic*. Aug. At http://ngm.nationalgeographic.com/2013/08/sugar/

cohen-text.
———. 2006. *Sweet and Low.* New York: Picador.
Coller, H. A. 2014. "Is Cancer a Metabolic Disease?" *American Journal of Pathology* 184, no. 1 (Jan.): 4–17.
Committee on Medical Aspects (COMA) of Food Policy. 1989. *Report on Health and Social Subjects:* no. 37, *Dietary Sugars and Human Disease.* London: Her Majesty's Stationery Office.
Coughlin, S. S., E. E. Calle, L. R. Teras, J. Petrelli, and M. J. Thun. 2004. "Diabetes Mellitus as a Predictor of Cancer Mortality in a Large Cohort of U.S. Adults." *American Journal of Epidemiology* 159, no. 12 (June 15): 1160–67.
Council on Foods and Nutrition (CFN). 1942. "Some Nutritional Aspects of Sugar, Candy and Sweetened Carbonated Beverages." *J.A.M.A.* 120, no. 10 (Nov. 7): 763–65.
Courtwright, D. 2001. *Forces of Habit: Drugs and the Making of the Modern World.* Cambridge, Mass.: Harvard University Press.
Coutinho, T. de A., S. T. Turner, P. A. Peyser, L. F. Bietak, P. F. Sheedy, and I. J. Kulloo. 2007. "Association of Serum Uric Acid with Markers of Inflammation, Met-

abolic Syndrome, and Subclinical Coronary Atherosclerosis." *American Journal of Hypertension* 20, no. 1 (Jan.): 83–89.

Cray, D. W. 1969. "Battle over Sweeteners Turns Bitter." *New York Times,* June 1: F12.

Cutting, W. C. 1943. "The Treatment of Obesity." *Clinical Endocrinology* 3, no. 2 (Feb.): 85–88.

Dabelea, D., W. C. Knowler, and D. J. Pettitt. 2000. "Effect of Diabetes in Pregnancy on Offspring: Follow-up Research in the Pima Indians." *Journal of Maternal-Fetal Medicine* 9, no. 1 (Jan.–Feb.): 83–88.

Dahl, R. 1984. *Boy: Tales of Childhood.* New York: Penguin.

Darby, W. J., C. G. Salsbury, W. J. McGanity, H. F. Johnson, E. B. Bridgforth, and H. R. Sandstead. 1956. "A Study of the Dietary Background and Nutriture of the Navajo Indian." Supplement, *Journal of Nutrition* 60, no. 2 (Nov.): 1–85.

Darrow, R. W., and D. J. Forrestal. 1979. *The Dartnell Public Relations Handbook,* 4th edition. Chicago: Dartnell Corporation.

Davies, L. E. 1950. "$4,000,000 Is Raised in Heart Campaign." *New York Times,* June 25: 37.

Dawber, T. R. 1978. "Annual Discourse–

Unproved Hypotheses." *New England Journal of Medicine* 299, no. 9 (Aug. 31): 452–58.

Deerr, N. 1950. *The History of Sugar.* Vol. 2. London: Chapman and Hall.

———. 1949. *The History of Sugar.* Vol. 1. London: Chapman and Hall.

DeFronzo, R. A. 1981. "The Effect of Insulin on Renal Sodium Metabolism: A Review with Clinical Implications." *Diabetologia* 21, no. 3 (Sept.): 165–71.

DePue, J. D., R. K. Rosen, M. Batts-Turner, et al. 2010. "Cultural Translation of Interventions: Diabetes Care in American Samoa." *American Journal of Public Health* 100, no. 10 (Nov.): 2085–93.

Després, J. P., B. Lamarche, P. Mauriège, et al. 1996. "Hyperinsulinemia as an Independent Risk Factor for Ischemic Heart Disease." *New England Journal of Medicine* 334, no. 15 (April 11): 952–57.

Deutsch, R. M. 1975. "Sugar in the Diet of Man: A Summary." Washington, D.C. Sugar Association, Inc., Records of the Great Western Sugar Company. Colorado Agricultural Archive, Colorado State University.

DGF. 1972. "Dr. John Hickson." Sept. 8. British American Tobacco. At https://industrydocuments.library.ucsf.edu/

tobacco/docs/#id=gjjy0205.
Dickson, J. A. S. 1964. "Dietary Fat and Dietary Sugar." *Lancet* 284 (Aug. 15): 361.
Dix, D. 1904. "Causes and Cure of Rheumatism." *New York Times,* Feb. 21: 32.
Doll, R., and A. B. Hill. 1964. "Mortality in Relation to Smoking: Ten Years' Observations of British Doctors." *British Medical Journal* 1 (May 30): 1399–1410.
Doll, R., and R. Peto. 1981. "The Causes of Cancer: Quantitative Estimates of Avoidable Risks of Cancer in the United States Today." *Journal of the National Cancer Institute* 66, no. 6 (June): 1191–1308.
Domino Sugar. 1953. *Life,* April 20: 116.
Donnison, J. P. 1929. "Blood Pressure in the African Native." *Lancet* 213 (Jan. 5): 6–7.
Drummond, J. C., and A. Wilbraham. 1994. *The Englishman's Food.* London: Pimlico.
Duffey, K. J., and B. M. Popkin. 2008. "High-Fructose Corn Syrup: Is This What's for Dinner?" *American Journal of Clinical Nutrition* 88 (supplement), no. 6 (Dec.): 1722S–32S.
Dukes, C. E. 1964. "The Origin and Early History of the Imperial Cancer Research Fund." *Annals of the Royal College of*

Surgeons of England 36 (June): 325–38.
Economic Research Service (ERS), United States Department of Agriculture. 2015. "Selected Fruit Juices: Per Capita Availability." At http://www.ers.usda.gov/datafiles/Food_Availabily_Per_Capita_Data_System/ Food_Availability/fruitju.xls.
Elgin, Bruce, V. A., Earl of. 1906. The Secretary of State to the Governors, &c.: no. 5. In *Further Correspondence Relating to the Cancer Research Scheme* (London: His Majesty's Stationery Office), 3–5. In Parliamentary Papers, House of Commons and Command, vol. 70.
Ellestad-Sayad, J. J., J. C. Haworth, and J. A. Hildes. 1978. "Disaccharide Malabsorption and Dietary Patterns in Two Canadian Eskimo Communities." *American Journal of Clinical Nutrition* 31, no. 8 (Aug.): 1473–78.
Elson, L.A., T. E. Betts, and R. D. Passey. 1972. "The Sugar Content and the pH of the Smoke of Cigarette, Cigar and Pipe Tobaccos in Relation to Lung Cancer." *International Journal of Cancer* 9, no. 3 (May): 666–75.
Emerson, H., and L. D. Larimore. 1924. "Diabetes Mellitus–A Contribution to Its Epidemiology Based Chiefly on Mortality

Statistics." *Archives of Internal Medicine* 34, no. 5 (Nov.): 585–630.

Emmerson, B. T. 1996. "The Management of Gout." *New England Journal of Medicine* 334, no. 7 (Feb. 15): 445–51.

Enos, W. F., R. H. Holmes, and J. Beyer. 1953. "Coronary Disease Among United States Soldiers Killed in Action in Korea: A Preliminary Report." *J.A.M.A.* 152, no. 12 (July 18): 1090–93.

Ervin, R. B. 2009. "Prevalence of Metabolic Syndrome Among Adults 20 Years of Age and Over, by Sex, Age, Race and Ethnicity, and Body Mass Index: United States, 2003–2006." *National Health Statistics Reports* no. 13 (May 5). At http://www.cdc.gov/nchs/data/nhsr/nhsr013.pdf.

Evans, J. M., L. A. Donnelly, A. M. Emslie-Smith, D. R. Alessi, and A. D. Morris. 2005. "Metformin and Reduced Risk of Cancer in Diabetic Patients." *British Medical Journal* 330 (June 4): 1304–5.

Ewen, S. 1998. "Leo Burnett: Sultan of Sell." *Time,* Dec. 7. At http://content.time.com/time/magazine/article/0,9171,989783,00.html.

Fam, A. G. 2002. "Gout, Diet, and the Insulin Resistance Syndrome." *Journal of Rheumatology* 29, no. 7 (July): 1350–55.

Feig, D. I., D.-H. Kang, and R. J. Johnson.

2008. "Uric Acid and Cardiovascular Risk." *New England Journal of Medicine* 359, no. 17 (Oct. 23): 1611–21.

Felita, L. S., E. Sobngwi, P. Serradas, F. Calvo, and J. F. Gautier. 2006. "Consequences of Fetal Exposure to Maternal Diabetes in Offspring." *Journal of Clinical Endocrinology and Metabolism* 91, no. 10 (Oct.): 3718–24.

Ferguson, N. 2002. *Empire: The Rise and Demise of the British World Order and the Lessons for Global Power.* London: Penguin.

Fernstrom, J. D., S. D. Munger, A. Sclafani, I. E. de Araujo, A. Roberts, and S. Molinary. 2012. "Mechanisms for Sweetness." *Journal of Nutrition* 142 (supplement), no. 6 (June): 1134S–41S.

Feudtner, C. 2003. *Bittersweet: Diabetes, Insulin and the Transformation of Illness.* Chapel Hill: University of North Carolina Press.

Fitz, R. H., and E. P. Joslin. 1898. "Diabetes Mellitus at the Massachusetts General Hospital from 1824 to 1898: A Study of the Medical Records." *J.A.M.A.* 31 (July 23): 165–71.

Flanagan, G. M. 1943. "Candy on Two Fronts." *New York Times,* Feb. 21: SM10.

Fleming, H. C. 1924. *Medical Observations*

on the Zuni Indians. New York: Museum of the American Indian Heye Foundation.

Flexner, A. 1910. *Medical Education in the United States and Canada.* New York: Carnegie Foundation.

Food and Agricultural Organization (FAO). n.d. "The Nutrition Transition and Obesity." At http://www.fao.org/focus/e/obesity/obes2.htm.

Fosdick, L. S. 1952. "Some New Concepts Concerning the Role of Sugar in Dental Caries." *Oral Surgery, Oral Medicine, and Oral Pathology* 5, no. 6 (June): 615–24.

Frantz, I. D., Jr., E. A. Dawson, P. L. Ashman, et al. 1989. "Test of Effect of Lipid Lowering by Diet on Cardiovascular Risk: The Minnesota Coronary Survey." *Arteriosclerosis* 9, no. 1 (Jan.–Feb.): 129–35.

Friedman, J. M. 2004. "Modern Science Versus the Stigma of Obesity." *Nature Medicine* 10, no. 6 (June): 563–69.

Frost, R. 1965. "Sugar Industry Eyes New Fields." *New York Times,* Jan. 3: 135.

Gale, E. A. M. 2013. "Commentary: The Hedgehog and the Fox: Sir Harold Himsworth (1905–93)." *International Journal of Epidemiology* 42, no. 6 (Dec.): 1602–7.

Galloway, J. H. 1989. *The Sugar Cane Industry: An Historical Geography from Its*

Origins to 1914. New York: Cambridge University Press.

Galton, L. 1976. *The Truth About Fiber in Your Food.* New York: Crown.

Gardner, H. W. 1901. "The Dietetic Value of Sugar." *British Medical Journal* 1 (April 27): 1010–13.

Garner, W. W. 1946. *The Production of Tobacco.* Philadelphia: Blakiston Company.

Geibel, E. 2010. "Why Me? Understanding the Causes of Diabetes." *Diabetes Forecast,* Oct. At http://www.diabetesforecast.org/2010/oct/why-me-understanding-the-causes-of-diabetes.html.

Genuth, S. M., P. H. Bennett, M. Miller, and T. A. Burch. 1967. "Hyperinsulinism in Obese Diabetic Pima Indians." *Metabolism* 16, no. 11 (Nov.): 1010–15.

Gertler, M. M., S. M. Garn, and S. A. Levine. 1951. "Serum Uric Acid in Relation to Age and Physique in Health and in Coronary Heart Disease." *Annals of Internal Medicine* 36, no. 6 (June): 1421–31.

Gibson, A. 1917. "The Case Against Sugar." *Medical Summary* 39 (Oct.): 237–39.

Giovannucci, E. 2001. "Insulin, Insulin-Like Growth Factors and Colon Cancer: A Review of the Evidence." Supplement, *Journal of Nutrition* 131, no. 11 (Nov.):

3109S–20S.
Glinsmann, W. H., H. Irausquin, and Y. K. Park. 1986. “Report from FDA’s Sugars Task Force, 1986: Evaluation of Health Aspects of Sugars Contained in Carbohydrate Sweeteners.” Supplement, *Journal of Nutrition* 116, no. 11 (Nov.): S1–S216.
Global Energy Balance Network (GEBN). 2015a. “Energy Balance Basics.” Formerly online, downloaded Oct. 24, 2015, at https://gebn.org/energy-balance-basics.
———. 2015b. “Why Join GEBN?” Formerly online, downloaded Oct. 24, 2015, at https://gebn.org/membership.
Gohdes, D.M. 1986. “Diabetes in American Indians: A Growing Problem.” *Diabetes Care* 9, no. 6 (Nov.–Dec.): 609–13.
Grafe, E. 1933. *Metabolic Diseases and Their Treatment.* Trans. M. G. Boise. Philadelphia: Lea & Febiger.
Grande, F. 1975. “Sugar and Cardiovascular Disease.” *World Review of Nutrition and Dietetics* 22: 248–69.
Grande, F., J. T. Anderson, and A. Keys. 1974. “Sucrose and Various Carbohydrate-Containing Foods and Serum Lipids in Man.” *American Journal of Clinical Nutrition* 27, no. 10 (Oct.): 1043–51.
Graudal, N. A., T. Hubeck-Graudal, and

G. Jurgens. 2011. "Effects of Low Sodium Versus High Sodium Diet on Blood Pressure, Renin, Aldosterone, Catecholamines, Cholesterol, and Triglyceride." *Cochrane Database of Systematic Reviews* no. 11 (Nov. 9): CD004022.

Greene, R., ed. 1951. *The Practice of Endocrinology.* Philadelphia: J. B. Lippincott.

Greenwood, M. R., M. Cleary, L. Steingrimsdottir, and J. R. Vaselli. 1981. "Adipose Tissue Metabolism and Genetic Obesity: The LPL hypothesis." In *Recent Advances in Obesity Research,* Vol. III, ed. P. Björntorp, M. Cairella, and A. N. Howard (London: John Libbey, 1981), 75–79.

Gregg, E. W., X. Zhou, Y. J. Cheng, A. L. Albright, K. M. Narayan, and T. J. Thompson. 2014. "Trends in Lifetime Risk and Years of Life Lost Due to Diabetes in the USA, 1895–2011: A Modeling Study." *Lancet Diabetes & Enocrinology* 2, no. 11 (Nov.): 867–74.

Guthrie, C. 2007. "Is Alzheimer's a Form of Diabetes?" *Time,* October 18. At http://content.time.com/time/health/article/0,8599,1673236,00.html.

Hamilton, A. 2009. *Squeezed: What You Don't Know About Orange Juice.* New Haven, Conn.: Yale University Press.

Handler, P. 1975. "Welcome." In NAS 1975, 3–5.

Hannah, A. C., and D. Spence. 1996. *The International Sugar Trade.* New York: John Wiley & Sons.

Harding, W. R., C. E. Russell, F. Davidson, and I. A. M. Prior. 1986. "Dietary Surveys from the Tokelau Island Migrant Study." *Ecology of Food and Nutrition* 19, no. 2: 83–97.

Hass, H. B. 1960. Letter to Roger Adams, April 29. Sugar Research Foundation, Inc. Papers of Roger Adams, University of Illinois Archives, University of Illinois at Urbana-Champaign.

He, F. J., J. Li, and G. A. MacGregor. 2013. "Effect of Longer Term Modest Salt Reduction on Blood Pressure: Cochrane Systematic Review and Meta-Analysis of Randomised Trials." *Cochrane Database of Systematic Reviews* no. 4 (April 30): CD004937.

Heinbecker, P. 1928. "Studies on the Metabolism of Eskimos." *Journal of Biological Chemistry* 80, no. 2 (Dec. 1): 461–75.

Helmchen, L. A., and R. M. Henderson. 2004. "Changes in the Distribution of Body Mass Index of White US Men, 1890–2000." *Annals of Human Biology* 31, no. 2 (March–April): 174–81.

Hess, J. L. 1978. "Harvard's Sugar-Pushing Nutritionist." *Saturday Review,* Aug.: 10–14.

Hess, J. L., and K. Hess. 2000. *The Taste of America.* Champaign: University of Illinois Press.

Hesse, F. G. 1959. "A Dietary Study of the Pima Indian." *American Journal of Clinical Nutrition* 7 (Sept.–Oct.): 532–37.

Hetherington, A. W., and S. W. Ranson. 1942. "The Spontaneous Activity and Food Intake of Rats with Hypothalamic Lesions." *American Journal of Physiology* 136, no. 4 (June): 609–17.

———. 1939. "Experimental Hypothalamico-Hypophyseal Obesity in the Rat." *Proceedings of the Society for Experimental Biology and Medicine* 41, no. 2 (June): 465–66.

Heusen, J. C., A. Coune, and R. Heimann. 1967. "Cell Proliferation Induced by Insulin Organ Culture of Rat Mammary Carcinoma." *Experimental Cell Research* 45, no. 2 (Feb.): 351–60.

Hickson, J. L. 1975. "Sucrochemistry: In Planning the Research Effort." Internal document, Sept. 11, 12. International Sugar Research Foundation, Inc., Records of the Great Western Sugar Company. Colorado Agricultural Archive, Colorado

State University.
———. 1962. Letter to the Scientific Advisory Board, Nov. 5. Sugar Research Foundation, Inc., Papers of Roger Adams, University of Illinois Archives, University of Illinois at Urbana-Champaign.
Higgins, H. L. 1916. "The Rapidity with Which Alcohol and Some Sugars May Serve as a Nutriment." *American Journal of Physiology* 41, no. 2 (Aug. 1): 258–65.
Higginson, J. 1997. "From Geographical Pathology to Environmental Carcinogenesis: A Historical Reminiscence." *Cancer Letters* 117, no. 2 (Aug. 19): 133–42.
———. 1983. "Developing Concepts on Environmental Cancer: The Role of Geographical Pathology." *Environmental Mutagenesis* 5, no. 6: 929–40.
———. 1981. "Rethinking the Environmental Causation of Human Cancer." *Food and Cosmetics Toxicology* 19, no. 5 (Oct.): 539–48.
Hildes, J. A., and O. Schaefer. 1984. "The Changing Picture of Neoplastic Disease in the Western and Central Canadian Arctic (1950–1980). *Canadian Medical Association Journal* 130, no. 1 (Jan. 1): 25–32.
Hillebrand, S. S., ed. 1974. *Is the Risk of*

Becoming Diabetic Affected by Sugar Consumption: Proceedings of the Eighth International Sugar Research Symposium. Washington, D.C.: International Sugar Research Foundation.

Himsworth, H. P. 1949a. "Diet in the Aetiology of Human Diabetes." *Proceedings of the Royal Society of Medicine* 42, no. 5 (May): 323–26.

———. 1949b. "The Syndrome of Diabetes Mellitus and Its Causes." *Lancet* 253, no. 6,551 (March 19): 465–73.

———. 1935. "Diet and the Incidence of Diabetes Mellitus." *Clinical Science* 2, no. 1 (Sept.): 117–48.

———. 1931a. "High Carbohydrate Diet in Diabetes." *Lancet* 218 (Nov. 14): 1103.

———. 1931b. "Recent Advances in the Treatment of Diabetes." *Lancet* 218 (Oct. 31): 978–79.

Hockett, R. C. 1947. "The Progress of Sugar Research." *Scientific Monthly* 65, no. 4 (Oct.): 269–82.

Hoffman, F. L. 1915. *The Mortality from Cancer Throughout the World.* Newark, N.J.: Prudential Press.

Hooper, L., N. Martin, A. Abdelhamid, and G. Davey Smith. 2015. "Reduction in Saturated Fat Intake for Cardiovascular Disease." *Cochrane Database of Systematic*

Reviews no. 6 (June 10): CD011737.
Hooper, L., C. D. Summerbell, R. Thompson R. et al. 2012. "Reduced or Modified Dietary Fat for Preventing Cardiovascular Disease." *Cochrane Database of Systematic Reviews* no. 5 (May 16): CD002137.
House Committee on Government Operations. 1970. *Cyclamate Sweeteners.* Hearing before a subcommittee of the Committee on Government Operations, House of Representatives, 91st Congress, June 10. Washington, D.C.: U.S. Government Printing Office.
Howard, B. V., J. E. Manson, M. L. Stefanick, et al. 2006. "Low-Fat Dietary Pattern and Weight Change over 7 Years: The Women's Health Initiative Dietary Modification Trial." *J.A.M.A.* 295, no. 1 (Jan. 4): 39–49.
Howard, B. V., L. Van Horn, J. Hsia, et al. 2006. "Low-Fat Dietary Pattern and Risk of Cardiovascular Disease: The Women's Health Initiative Randomized Controlled Dietary Modification Trial." *J.A.M.A.* 295, no. 6 (Feb. 8): 655–66.
Hrdlička, A. 1908. *Physiological and Medical Observations Among the Indians of Southwestern United States and Northern Mexico.* Washington, D.C.: U.S. Government Printing Office.

———. 1906. "Notes on the Pima of Arizona." *American Anthropologist* 8, no. 1 (Jan.–Mar.): 39–46.

Hudson, E. H., and A. L. Young. 1931. "Medical and Surgical Practice on the Euphrates River: An Analysis of Two Thousand Consecutive Cases at Deir-Ez-Zor, Syria." *American Journal of Tropical Medicine* 11, no. 4 (July): 297–310.

Huetz de Lemps, A. 1999. "Colonial Beverages and the Consumption of Sugar." In *Food: A Culinary History from Antiquity to the Present,* ed. J.-L. Flandrin and M. Montanari (New York: Penguin), 383–93.

Hulthe, J., L. Bokemark, J. Wikstrand, and B. Fagerberg. 2000. "The Metabolic Syndrome, LDL Particle Size, and Atherosclerosis: The Atherosclerosis and Insulin Resistance Study." *Arteriosclerosis, Thrombosis, and Vascular Biology* 20, no. 9 (Sept.): 2140–47.

Huntsman, J., and A. Hooper. 1996. *Tokelau: A Historical Ethnography.* Auckland, N.Z.: Auckland University Press.

Hurd, M. D., P. Martorell, A. Delavanda, K. J. Mullen, and K. M. Langa. 2013. "Monetary Costs of Dementia in the United States." *New England Journal of*

Medicine 368, no. 14 (April 4): 1326–34.
Hydrick, C. R., and I. H. Fox. 1984. "Nutrition and Gout." In *Present Knowledge in Nutrition,* 5th edition, ed. R. E. Olson, H. P. Broquist, C. O. Chichester, W. J. Darby, A. C. Kolbye, Jr., and R. M. Stalvey (Washington, D.C.: Nutrition Foundation, 1984), 740–56.
Ingram, J. 2015. *The End of Memory.* New York: St. Martin's Press.
Institute of Medicine (IOM) of the National Academies. 2005. *Dietary Reference Intakes: Energy, Carbohydrate, Fiber, Fat, Fatty Acids, Cholesterol, Protein, and Amino Acids.* Washington D.C.: National Academies Press.
Insull, W., Jr., T. Oiso, and K. Tsuchiya. 1968. "Diet and Nutritional Status of Japanese." *American Journal of Clinical Nutrition* 21, no. 7 (July): 753–77.
International Diabetes Federation (IDF). 2015. "Diabetes Atlas," 7th edition. At http://www.idf.org/diabetesatlas.
———. 2013. *Diabetes Atlas,* 6th edition. At http://www.idf.org/diabetesatlas.
International Sugar Research Foundation (ISRF). 1976. Memo to members, April 30. "Developments in Brief: ISRF Support of Health Research and International Symposia." Washington, D.C.. Internal

document, Sugar Association, Inc., Records of the Great Western Sugar Company, Colorado Agricultural Archive, Colorado State University.

———. 1975. "Planning the Research Effort." Bethesda, Md. Internal document, International Sugar Research Foundation, Inc., Records of the Great Western Sugar Company. Colorado Agricultural Archive, Colorado State University.

———. 1969. Minutes of meeting of the Scientific Advisory Board, Dec. 5, 1969, Washington, D.C. International Sugar Research Foundation, Inc. Papers of Roger Adams, University of Illinois Archives, University of Illinois at Urbana-Champaign.

Intersalt Cooperative Research Group. 1988. "Intersalt, an International Study of Electrolyte Excretion and Blood Pressure: Results for 24 Hour Urinary Sodium and Potassium Excretion. *British Medical Journal* 297 (July 30): 319–28.

Inter-Society Commission for Heart Disease Resources. 1970. "Report of Inter-Society Commission for Heart Disease Resources: Prevention of Cardiovascular Disease. Primary Prevention of the Atherosclerotic Diseases." *Circulation* 42, no. 6 (Dec.): A55–A95.

Jacobson, M. 1978. "The Deadly White Powder." *Mother Jones* (July): 12–20.

Jenkins, D. J., T. M. Wolever, R. H. Taylor, et al. 1981. "Glycemic Index of Foods: A Physiological Basis for Carbohydrate Exchange." *American Journal of Clinical Nutrition* 34, no. 3 (March): 362–66.

Joe, J. R., and R. S. Young, eds. 1994. *Diabetes as a Disease of Civilization: The Impact of Culture Change on Indigenous Peoples.* New York: Mouton de Gruyter.

Johnson, R. J., J. Herrera-Acosta, G. F. Schreiner, and B. Rodriguez-Iturbe. 2002. "Subtle Acquired Renal Injury as a Mechanism of Salt-Sensitive Hypertension." *New England Journal of Medicine* 346, no. 12 (March 21): 913–23.

Johnson, R. J., M. S. Segal, T. Nakagawa, et al. 2007. "Potential Role of Sugar (Fructose) in the Epidemic of Hypertension, Obesity and the Metabolic Syndrome, Diabetes, Kidney Disease, and Cardiovascular Disease." *American Journal of Clinical Nutrition* 86, no. 4 (Oct.): 899–906.

Jordão, A. 1867. "On Some Symptoms of Diabetes: A Clinical Lecture Delivered in the Medical School of Lisbon." *American Journal of the Medical Sciences* 53, no. 106 (April): 510–12.

———. 1866. "Studies on Diabetes." *American Journal of the Medical Sciences* 54, no. 104 (Oct.): 467–80.

Jørgensen, M. E., K. Borch-Johnsen, D. R. Witte, and P. Bjerregaard. 2012. "Diabetes in Greenland and Its Relationship with Urbanization." *Diabetic Medicine* 29, no. 6 (June): 755–60.

Joslin, E. P. 1950. "A Half Century's Experience in Diabetes Mellitus." *British Medical Journal* 1 (May 13): 1095–98.

———. 1940. "The Universality of Diabetes." *J.A.M.A.* 115, no. 24 (Dec. 14): 2033–38.

———. 1934. "Studies in Diabetes Mellitus. II: Its Incidence and the Factors Underlying Its Variations." *American Journal of Medical Science* 187, no. 4 (April): 433–57.

———. 1928. *The Treatment of Diabetes Mellitus,* 4th edition. Philadelphia: Lea & Febiger.

———. 1927. "Arteriosclerosis and Diabetes." *Annals of Clinical Medicine* 5, no. 12: 1061–79.

———. 1923. *The Treatment of Diabetes Mellitus,* 3rd edition. Philadelphia: Lea & Febiger.

———. 1921. "The Prevention of Diabetes

Mellitus." *J.A.M.A.* 76, no. 2 (Jan. 8): 79–84.
———. 1917. *The Treatment of Diabetes Mellitus,* 2nd edition. Philadelphia: Lea & Febiger.
———. 1916. *The Treatment of Diabetes Mellitus.* Philadelphia: Lea & Febiger.
Joslin, E. P., L. I. Dublin, and H. H. Marks. 1934. "Studies in Diabetes Mellitus. II: Its Incidence and the Factors Underlying Its Variations." *American Journal of the Medical Sciences* 187, no. 4 (April): 433–57.
Justice, J. W. 1994. "The History of Diabetes Mellitus in the Desert People." In Joe and Young, eds., 1994, 69–127.
Kaaks, R. 1996. "Nutrition, Hormones, and Breast Cancer: Is Insulin the Missing Link?" *Cancer Causes and Control* 7, no. 6 (Nov.): 605–25.
Kaaks, R., and A. Lukanova. 2001. "Energy Balance and Cancer: The Role of Insulin and Insulin-Like Growth Factor I." *Proceedings of the Nutrition Society* 60, no. 1 (Feb.): 91–106.
Kahn, C. R., G. C. Weir, G. L. King, A. M. Jacobson, A. C. Moses, and R. J. Smith, eds. 2005. *Joslin's Diabetes Mellitus,* 14th edition. New York: Lippin-

cott, Williams & Wilkins.
Kare, M. R. 1975. "Monellin." In NAS 1975, 196–206.
Karolinska Institute. 1977. "Press Release: The 1977 Nobel Prize in Physiology or Medicine." At http://nobelprize.org/nobel_prizes/medicine/laureates/1977/press.html.
Kean, B. H. 1944. "The Blood Pressure of the Cuna Indians." *American Journal of Tropical Medicine* 24, no. 6 (Nov.): 341–43.
Kean, B. H., and J. F. Hammill. 1949. "Anthropathology of Arterial Tension." *Annals of Internal Medicine* 83, no. 3 (March 1): 355–62.
Kearns, C. E., S. A. Glantz, and L. A. Schmidt. 2015. "Sugar Industry Influence on the Scientific Agenda of the National Institute of Dental Research's 1971 National Caries Program: A Historical Analysis of Internal Documents." *PLOS Medicine* 12, no. 3 (March 10): e1001798.
Kelly, N. 1969. "What's at Stake in Sugar Research?" Bethesda, Md. International Sugar Research Foundation, Inc., Papers of Roger Adams, University of Illinois Archives, University of Illinois at Urbana-Champaign.

Keys, A. 1971. "Sucrose in the Diet and Coronary Heart Disease." *Atherosclerosis* 14, no. 2 (Sept.–Oct.): 193–202.

Keys, A., and M. Keys. 1975. *How to Eat Well and Stay Well the Mediterranean Way.* Garden City, N.Y.: Doubleday.

Khardori, R. 2015. "Diabetes Mellitus Medication." Medscape. At http://emedicine.medscape.com/article/117853-medication.

Kinugasa, A., K. Tsunamoto, N. Furukawa, T. Sawada, T. Kusunoki, and N. Shimada. 1984. "Fatty Liver and Its Fibrous Changes Found in Simple Obesity of Children." *Journal of Pediatric Gastroenterology and Nutrition* 3, no. 3 (June): 408–14.

Kleinridders, A., H. A. Ferris, W. Cai, and C. R. Kahn. 2014. "Insulin Action in Brain Regulates Systematic Metabolism and Brain Function." *Diabetes* 63, no. 7 (July): 2232–43.

Kohn, L. A., S. A. Levine, and M. Matton. 1925. "Sugar Content of the Blood in Runners Following a Marathon Race." *J.A.M.A.* 85, no. 7 (Aug. 15): 508–9.

Kolata, G. 1987. "High-Carb Diets Questioned." *Science* 235 (Jan. 9): 164.

Koop, C. E. 1988. "Message from the Surgeon General." In US HHS 1988.

Kotchen, K.A. 2011. "Historical Trends and Milestones in Hypertension Research: A Model of the Process of Translational Research." *Hypertension* 58, no. 4 (Oct.): 522–38.

Kraus, B. R., and B. M. Jones. 1954. *Indian Health in Arizona: A Study of Health Conditions Among Central and Southern Arizona Indians.* Tucson: University of Arizona Press.

Krauss, E. A. 1947. "Soft Drink Industry on Eve of New Growth Phase." *Forbes* 59, no. 12 (June 5): 36–37.

Kraybill, H. F. 1975. "The Question of Benefits and Risks." In NAS 1975, 59–75.

Krebs, H. A. 1967. "The Making of a Scientist." *Nature* 215, no. 5,109 (Sept. 30): 1441–45.

Kretchmer, N., and C. B. Hollenbeck, eds. 1991. *Sugars and Sweeteners.* Boca Raton, Fla.: CRC Press.

Laffer, C. L., and F. Elijovich. 2013. "Differential Predictors of Insulin Resistance in Nondiabetic Salt-Resistant and Salt-Sensitive Subjects." *Hypertension* 61, no. 3 (March): 707–15.

Lamborn, O. 1942. "A Suggested Program for the Cane and Beet Sugar Industries." Unpublished sugar-industry document, New York. Braga Brothers Collection,

Special and Area Studies Collections, George A. Smathers Libraries, University of Florida, Gainesville, Fla. At http://www.motherjones.com/documents/480900-a-suggested-program-for-the-cane-and-beet-sugar#document/p1/a79758.

Landa, M. M. 2012. "Response to Petition from Corn Refiners Association to Authorize 'Corn Sugar' as an Alternate Common or Usual Name for High Fructose Corn Syrup (HFCS)." Letter to Audrae Erickson. At http://www.fda.gov/aboutFDA/CentersOffices/OfficeofFoods/CFSAN/CFSANFOIAElectronicReadingRoom/ucm305226.htm.

Landsberg, L. 2001. "Insulin-Mediated Sympathetic Stimulation: Role in the Pathogenesis of Obesity-Related Hypertension (or, How Insulin Affects Blood Pressure, and Why)." *Journal of Hypertension* 19, no. 3, pt. 2 (March 19): 523–28.

———. 1986. "Diet, Obesity and Hypertension: An Hypothesis Involving Insulin, the Sympathetic Nervous System, and Adaptive Thermogenesis." *Quarterly Journal of Medicine* 61, no. 236 (Dec.): 1081–90.

Lastra, G., S. Dhuper, M. S. Johnson, and J. R. Sowers. 2010. "Salt, Aldosterone, and Insulin Resistance: Impact on the Cardiovascular System." *Nature Reviews: Cardiol-*

ogy 7, no. 10 (Oct.): 577–84.
Lawrence, J. S., T. Behrend, P. H. Bennett, et al. 1966. "Geographical Studies on Rheumatoid Arthritis." *Annals of Rheumatoid Arthritis* 25, no. 5 (Sept.): 425–32.
Lawrie, M. 1928. "Nervous Instability and the Intake of Sugar." *Lancet* 211 (Jan. 21): 158.
Lee, M. O., and N. K. Schaffer. 1934. "Anterior Pituitary Growth Hormone and the Composition of Growth." *Journal of Nutrition* 7, no. 3 (March): 337–63.
Leibson, C. L., W. A. Rocca, V. A. Hanson, et al. 1997. "Risk of Dementia Among Persons with Diabetes Mellitus: A Population-Based Cohort Study." *American Journal of Epidemiology* 145, no. 4 (Feb. 15): 301–8.
LeRoith, D., and C. T. Roberts, Jr. 2003. "The Insulin-Like Growth Factor System and Cancer." *Cancer Letters* 195, no. 2 (June 10): 127–37.
Levenstein, H. 1993. *Paradox of Plenty: A Social History of Eating in Modern America.* New York: Oxford University Press.
Levin, I. 1910. "Cancer Among the North American Indians and Its Bearing Upon the Ethnological Distribution of Disease." *Zeitschrift für Krebsforschung* 9, no. 3

(Oct.): 422–35.

Levitsky, D. A., I. Faust, and M. Glassman. 1976. "The Ingestion of Food and the Recovery of Body Weight Following Fasting in the Naive Rat." *Physiology & Behavior* 17, no. 4 (Oct.): 575–80.

Li, X., D. Song, and S. X. Leng. 2015. "Link Between Type 2 Diabetes and Alzheimer's Disease: From Epidemiology to Mechanism and Treatment." *Clinical Interventions in Aging* 10 (March 10): 549–60.

Life Sciences Research Office (LSRO). 1977. "The Public Responsibility of Scientific Societies." *Federation Proceedings* 36, no. 11 (Oct.): 2463–64.

———. 1976. *Evaluation of the Health Aspects of Sucrose as a Food Ingredient.* Bethesda, Md.: Federation of American Societies for Experimental Biology.

———. 1975. *Tentative Evaluation of the Health Aspects of Sucrose as a Food Ingredient.* Bethesda, Md.: Federation of American Societies for Experimental Biology.

Lipid Research Clinics (LRC) Program. 1984a. "The Lipid Research Clinics Coronary Primary Prevention Trial Results: I, Reduction in Incidence of Coronary Heart Disease." *J.A.M.A.* 251, no. 3 (Jan. 20): 351–64.

——. 1984b. "The Lipid Research Clinics Coronary Primary Prevention Trial Results: II, The Relationship of Reduction in Incidence of Coronary Heart Disease to Cholesterol Lowering." *J.A.M.A.* 251, no. 3 (Jan. 20): 365–74.

Long, C. N. H. 1927. "Etiology." In *Diseases of Metabolism,* 2nd edition, ed. G. G. Duncan (Philadelphia: W. B. Saunders, 1927), 710–20.

Lovegren, S. 2012. "Breakfast Foods." In *The Oxford Encyclopedia of Food and Drink in America,* 2nd edition, ed. A. Smith (Oxford, U.K.: Oxford University Press, 2012), 207–15.

Lowenstein, F. W. 1961. "Blood-Pressure in Relation to Age and Sex in the Tropics and Subtropics: A Review of the Literature and an Investigation in Two Tribes of Brazil Indians." *Lancet* 277 (Feb. 18): 389–92.

———. 1954. "Some Epidemiologic Aspects of Blood Pressure and Its Relationship to Diet and Constitution with Particular Consideration of the Chinese: A Review of the Pertinent Literature of the Past 40 Years." *American Heart Journal* 47, no. 5 (June): 874–86.

Ludmerer, K. M. 1988. *Learning to Heal: The Development of American Medical Edu-*

cation. New York: Perseus Books.
Ludwig, J., T. R. Viggiano, D. B. McGill, and B. J. Oh. 1980. "Nonalcoholic Steatohepatitis: Mayo Clinic Experiences with a Hitherto Unnamed Disease." *Mayo Clinic Proceedings* 55, no. 7 (July): 434–38.
Lusk, G. 1933. *Nutrition.* New York: Paul B. Hoeber.
Luzardo, L., O. Noboa, and J. Boggia. 2015. "Mechanisms of Salt-Sensitive Hypertension." *Current Hypertension Reviews* 11, no. 1: 14–21.
Lyons, R. D. 1977. "F.D.A. Banning Saccharin Use on Cancer Links." *New York Times,* March 10: 1.
Lyssiotis, C. A., and L. C. Cantley. 2013. "F Stands for Fructose and Fat." *Nature* 502 (Oct. 10): 181–82.
Mann, C. C. 2011. *1493: Uncovering the New World Columbus Created.* New York: Knopf.
Marble, A., P. White, R. F. Bradley, and L. P. Krall, eds. 1971. *Joslin's Diabetes Mellitus,* 11th edition. Philadelphia: Lea & Febiger.
Marks, S. V., and K. E. Maskus. 1993. "Introduction." In Marks and Maskus, eds., 1993, 1–14.
Marks, S. V., and K. E. Maskus, eds. 1993.

The Economics and Politics of World Sugar Prices. Ann Arbor: University of Michigan Press.

Marmot, M. G., and S. L. Syme. 1976. "Acculturation and Coronary Heart Disease in Japanese-Americans." *American Journal of Epidemiology* 104, no. 3 (Sept.): 225–47.

Marshall, E. 1990. "Third Strike for NCI Breast Cancer Study." *Science* 250 (Dec. 14): 1503–4.

Masironi, R. 1970. "Dietary Factors and Coronary Heart Disease." *Bulletin of the World Health Organization* 42, no. 1: 103–14.

Mau, M. K., J. Keawe'aimoku Kaholokula, J. West, et al. 2010. "Translating Diabetes Prevention into Native Hawaiian and Pacific Islander Communities: The PILI 'Ohana Pilot Project." *Progress in Community Health Partnerships* 4, no. 1 (Spring): 7–16.

Mayer, J. 1976. "The Bitter Truth About Sugar." *New York Times,* June 20: 177.

———. 1968. *Overweight: Causes, Cost, and Control.* Englewood Cliffs, N.J.: Prentice-Hall.

———. 1953a. "Glucostatic Mechanism of Regulation of Food Intake." *New England*

Journal of Medicine 249, no. 1 (July 2): 13–16.
———. 1953b. "Decreased Activity and Energy Balance in the Hereditary Obesity-Diabetes Syndrome of Mice." *Science* 117 (May 8): 504–5.
Mayer, J., and J. Goldberg. 1986. "Signs of Atherosclerosis Show Up at an Early Age in Heart-Disease Study." *Chicago Tribune,* March 13.
Mayes, P. A. 1993. "Intermediary Metabolism of Fructose." Supplement, *American Journal of Clinical Nutrition* 58, no. 5 (Nov.): S754–S765.
McGandy, R. B., and J. Mayer. 1973. "Atherosclerotic Disease, Diabetes, and Hypertension: Background Considerations." In *U.S. Nutrition Policies in the Seventies,* ed. J. Mayer (San Francisco: W. H. Freeman), 37–43.
McGovern, G. 1977. Letter to J. W. Tatem, Jr., July 1, Records of the Great Western Sugar Company. Colorado Agricultural Archive, Colorado State University.
Menke, A., S. Casagrande, L. Geiss, and C. C. Cowie. 2015. "Prevalence of and Trends in Diabetes Among Adults in the United States, 1988–2012." *J.A.M.A.* 314, no. 10 (Sept. 8): 1021–29.
Miller, J. H., and M. D. Bogdonoff. 1954.

"Antidiuresis Associated with Administration of Insulin." *Journal of Applied Physiology* 6, no. 8 (Feb.): 509–12.

Miller, M., T. A. Burch, P. H. Bennett, and A. G. Steinberg. 1965. "Prevalence of Diabetes Mellitus in the American Indians: Results of Glucose Tolerance Tests in the Pima Indians of Arizona." *Diabetes* 14, no. 7 (July): 439–40.

Mills, C. A. 1930. "Diabetes Mellitus: Sugar Consumption in Its Etiology." *Archives of Internal Medicine* 46 (Oct.): 582–84.

Mintz, S. W. 1991. "Pleasure, Profit, and Satiation." In *Seeds of Change,* ed. J. J. Viola and C. Margolis (Washington, D.C.: Smithsonian Institution), 112–29.

———. 1985. *Sweetness and Power: The Place of Sugar in Modern History.* New York: Penguin.

Mitchell, H. S. 1930. "Nutrition Survey in Labrador and Northern Newfoundland." *Journal of the American Dietetic Association* 6, no. 1 (June): 29–35.

Moffat, R. U. 1904. Principal Medical Officer, Nairobi, to His Majesty's Commissioner. In Anon. 1906, 35–36.

Monod, J. 1965. "Nobel Lecture: From Enzymatic Adaption to Allosteric Transitions." Nobelprize.org. At http://www.nobelprize.org/nobel_prizes/medicine/

laureates/1965/monod-lecture.html.
Montanari, M. 1994. *The Culture of Food.* Trans. C. Ipsen. Oxford, U.K.: Blackwell.
Moore, J. S. 1890. "The Tax on Sugar: How the Many Are Fleeced for the Sake of a Few." Letter, *New York Times,* Feb. 19: 9.
Moore, W. W. 1983. *Fighting for Life: The Story of the American Heart Association, 1911–1975.* New York: New York Heart Association.
Moseley, B. 1799. *A Treatise on Sugar.* London: G. G. and J. Robinson.
Mouratoff, G. J., N. V. Carroll, and E. M. Scott. 1967. "Diabetes Mellitus in Eskimos." *J.A.M.A.* 199, no. 3 (March 27): 107–12.
Mouratoff, G. J., and E. M. Scott. 1973. "Diabetes Mellitus in Eskimos After a Decade." *J.A.M.A.* 226, no. 11 (Dec. 10): 1345–46.
Mrosovsky, N. 1976. "Lipid Programmes and Life Strategies in Hibernators." *American Zoologist* 16, no. 4 (Autumn): 685–97.
Multiple Risk Factor Intervention Trial Research Group (MRFIT). 1982. "Multiple Risk Factor Intervention Trial: Risk Factor Changes and Mortality Results." *J.A.M.A.* 248, no. 12 (Sept. 24): 1465–77.

Nagle, J. J. 1965. "Soft-Drink Brands Add and Multiply." *New York Times,* July 18: 122.

———. 1963. "Cola Producers Enter New Field." *New York Times,* May 19: 218.

National Academy of Sciences (NAS). 1975. *Sweeteners: Issues and Uncertainties.* Washington, D.C.: National Academy of Sciences.

National Analysts, Inc. 1974. "Attitudes Toward Sugar: A Study Conducted for the Sugar Association and the International Sugar Research Foundation." Records of the Great Western Sugar Company. Colorado Agricultural Archive, Colorado State University.

National Cancer Institute (NCI). 2009. "Artificial Sweeteners and Cancer." At http://www.cancer.gov/about-cancer/causes-prevention/risk/diet/artificial-sweeteners-fact-sheet.

National Commission on Diabetes. 1976. *Report of the National Commission on Diabetes to the Congress of the United States:* vol. III, *Reports of Committees, Subcommittees, and Workgroups.* HEW Publication No. (NIH)76-1021. Bethesda, Md.: Department of Health, Education, and Welfare.

National Heart, Lung, and Blood Institute

(NHLBI). 2015. "What Is Metabolic Syndrome?" At http://www.nhlbi.nih.gov/health/health-topics/topics/ms.

National Heart, Lung, and Blood Institute (NHLBI) Communication Office. 2006. "News from the Women's Health Initiative: Reducing Total Fat Intake May Have Small Effect on Risk of Breast Cancer, No Effect on Risk of Colorectal Cancer, Heart Disease, or Stroke." Press release, Feb. 7.

National Institute of Diabetes and Digestive and Kidney Diseases (NIDDK). 2014a. "Insulin Resistance and Prediabetes." At http://www.niddk.nih.gov/health-information/health-topics/Diabetes/insulin-resistance-prediabetes/Pages/index.aspx.

———. 2014b. "Nonalcoholic Steatohepatitis." At http://www.niddk.nih.gov/health-information/health-topics/liver-disease/nonalcoholic-steatohepatitis/Pages/facts.aspx.

———. 2012. "Overweight and Obesity Statistics." At http://www.niddk.nih.gov/health-information/health-statistics/Pages/overweight-obesity-statistics.aspx#a.

———. 2011. *Advances and Emerging Opportunities in Diabetes Research: A Strategic Planning Report of the Diabetes Mellitus Interagency Coordinating Committee.* NIH publication no. 11-7572. Bethesda, Md.: National Institutes of Health.

National Research Council (NRC), Committee on Diet and Health, Food and Nutrition Board, Commission on Life Sciences. 1989. *Diet and Health: Implications for Reducing Chronic Disease Risk.* Washington, D.C.: National Academies Press.

Nees, P. O., and P. H. Derse. 1965. "Feeding and Reproduction of Rats Fed Calcium Cyclamate." *Nature* 206, no. 5,005 (Oct. 2): 81–82.

Newburgh, L. H. 1942. "Obesity." *Archives of Internal Medicine* 70, no. 6 (Dec.): 1033–96.

Newburgh, L. H., and M. W. Johnston. 1930a. "The Nature of Obesity." *Journal of Clinical Investigation* 8, no. 2 (Feb.): 197–213.

———. 1930b. "Endogenous Obesity–A Misconception." *Annals of Internal Medicine* 8, no. 3 (Feb.): 815–25.

Newcombe, D. S. 2013. *Gout.* London: Springer-Verlag.

Nikkilä, E. A. 1974. "Influence of Dietary Fructose and Sucrose on Serum Triglycerides in Hypertriglyceridemia and Diabetes." In Sipple and McNutt, eds., 1974, 441–50.

Noorden, C. von. 1907. "Obesity." Trans. D. Spence. In *Metabolism and Practical Medicine,* Vol. 3: *The Pathology of Metabolism,* ed. C. von Noorden and I. W. Hall (Chicago: W. Keener, 1907), 693–715.

Noto, H., A. Goto, T. Tsujimoto, and M. Noda. 2012. "Cancer Risk in Diabetic Patients Treated with Metformin: A Systematic Review and Meta-Analysis." *PLOS One* 7, no. 3 (March): e33411.

Nuccio, S. 1964. "Advertising: Sales Clicking for Dietetic Pop." *New York Times,* May 20: 68.

O'Connor, A. 2015. "Coca-Cola Funds Effort to Alter Obesity Battle." *New York Times,* Aug. 10: A1.

Ors, R., E. Ozek, G. Baysoy, et al. 1999. "Comparison of Sucrose and Human Milk on Pain Response in Newborns." *European Journal of Pediatrics* 158, no. 1 (Jan.): 63–66.

Orwell, G. 1958. *Road to Wigan Pier.* New York: Harcourt. [Originally published in 1937.]

Osborne, C. K., G. Bolan, M. E. Monaco, and M. E. Lippman. 1976. "Hormone Responsive Human Breast Cancer in Long-Term Tissue Culture: Effect of Insulin." *Proceedings of the National Acad-*

emy of Sciences 73, no. 12 (Dec.): 4536–40.

Oscai, L. B., M. M. Brown, and W. C. Miller. 1984. "Effect of Dietary Fat on Food Intake, Growth and Body Composition in Rats." *Growth* 48, no. 4 (Winter): 415–24.

Osler, W. 1909. *The Principles and Practice of Medicine,* 7th edition. New York: D. Appleton.

———. 1901. *The Principles and Practice of Medicine,* 4th edition. New York: D. Appleton.

———. 1892. *The Principles and Practice of Medicine.* New York: D. Appleton.

Østbye, T., T. J. Welby, I. A. M. Prior, C. E. Salmond, and Y. M. Stokes. 1989. "Type 2 (Non-Insulin-Dependent) Diabetes Mellitus, Migration and Westernization: The Tokelau Island Migrant Study." *Diabetologia* 32, no. 8 (Aug.): 585–90.

Ott, A., R. P. Stolk, F. van Harskamp, H. A. Pols, A. Hofman, and M. M. Breteler. 1999. "Diabetes Mellitus and the Risk of Dementia: The Rotterdam Study." *Neurology* 53, no. 9 (Dec.): 1937–42.

Ott, A., R. P. Stolk, A. Hofman, F. van Harskamp, D. E. Grobbee, and M. M. Breteler. 1996. "Association of Diabetes Mellitus and Dementia: The Rotterdam Study." *Diabetologia* 39, no. 11 (Nov.):

1392–97.
Page, I. H., F. J. Stare, A. C. Corcoran, H. Pollack, and C. F. Wilkinson, Jr. 1957. "Atherosclerosis and the Fat Content of the Diet." *Circulation* 16, no. 2 (Aug.): 163–78.
Page, L. B., A. Damon, and R. C. Moellering, Jr. 1974. "Antecedents of Cardiovascular Disease in Six Solomon Islands Societies." *Circulation* 49, no. 6 (June): 1132–46.
Parks, J. and E. Waskow, 1961. "Diabetes Among the Pima Indians of Arizona." *Arizona Medicine* 18, no. 4 (April): 99–106.
PBS NewsHour. 2010. "Michelle Obama: Team Effort Needed to Halt Childhood Obesity." Feb. 9. At http://www.pbs.org/newshour/bb/health-jan-june10-firstlady_02-09.
Pendergrast, M. 1993. *For God, Country and Coca Cola.* New York: Basic Books.
Pennington, N. L., and C. W. Baker. 1990. *Sugar: A User's Guide to Sucrose.* New York: Van Nostrand Reinhold.
Perheentupa, J., and K. Raivio. 1967. "Fructose-Induced Hyperuricaemia." *Lancet* 290 (Sept.): 528–31.
Pettitt, D. J., K. A. Aleck, H. R. Baird, M. J. Carraher, P. H. Bennett, and W. C. Knowler. 1988. "Congenital Susceptibility

to NIDDM: Role of Intrauterine Environment." *Diabetes* 37, no. 5 (May): 622–28.

Pettitt, D. J., H. R. Baird, K. A. Aleck, P. H. Bennett, and W. C. Knowler. 1983. "Excessive Obesity in Offspring of Pima Indian Women with Diabetes During Pregnancy." *New England Journal of Medicine* 308, no. 5 (Feb.): 242–45.

Phillips, W. D., Jr. 1985. *Slavery from Roman Times to the Early Transatlantic Trade.* Manchester, U.K.: Manchester University Press.

Plice, M. J. 1952. "Sugar Versus the Intuitive Choice of Foods by Livestock." *Journal of Range Management* 5, no. 2 (March): 69–75.

Pollak, M. N., E. S. Schernhammer, and S. E. Hankinson. 2004. "Insulin-Like Growth Factors and Neoplasia." *Nature Reviews of Cancer* 4, no. 7 (July): 505–18.

Pollan, M. 2008. *In Defense of Food.* New York: Penguin.

———. 2002. "When a Crop Becomes King." *New York Times Magazine,* July 19: A17.

———. 2001. *The Botany of Desire: A Plant's-Eye View of the World.* New York: Random House.

Poloz, Y., and V. Stambolic. 2015. "Obesity

and Cancer: A Case for Insulin Signaling." *Cell Death and Disease* 6, no. 12 (Dec. 31): e2037.
Popper, K. R. 1979. *Objective Knowledge: An Evolutionary Approach.* Revised edition. Oxford, U.K.: Clarendon Press.
Porter, R., and G. S. Rousseau. 1998. *Gout: The Patrician Malady.* New Haven, Conn.: Yale University Press.
Prentice, R. L., B. Caan, R. T. Chlebowski, et al. 2006. "Low-Fat Dietary Pattern and Risk of Invasive Breast Cancer: The Women's Health Initiative Randomized Controlled Dietary Modification Trial." *J.A.M.A.* 295, no. 6 (Feb. 8): 629–42.
Presley, J. W. 1991. "A History of Diabetes Mellitus in the United States, 1880–1990." Ph.D. dissertation. University of Texas at Austin.
Price, R. A., M. A. Charles, D. J. Pettitt, and W. C. Knowler. 1993. "Obesity in Pima Indians: Large Increases Among Post–World War II Birth Cohorts." *American Journal of Physical Anthropology* 92, no. 4 (Dec.): 473–79.
Price, W. A. 1939. *Nutrition and Physical Degeneration.* New York: Paul B. Hoeber.
Priebe, P. M., and G. B. Kauffman. 1980. "Making Governmental Policy Under

Conditions of Scientific Uncertainty: A Century of Controversy About Saccharin in Congress and the Laboratory." *Minerva* 18, no. 4 (Winter): 556–74.

Prinsen Geerligs, H. C. 2010. *The World's Cane Sugar Industry, Past and Present.* Cambridge, U.K.: Cambridge University Press. [Originally published in 1912.]

Prior, I. A. 1971. "The Price of Civilization." *Nutrition Today,* July/Aug.: 2–11.

Prior, I. A., R. Beaglehole, F. Davidson, and C. E. Salmond. 1978. "The Relationships of Diabetes, Blood Lipids, and Uric Acid Levels in Polynesians." *Advances in Metabolic Disorders* 9: 241–61.

Prior, I. A., B. S. Rose, and F. Davidson. 1964. "Metabolic Maladies in New Zealand Maoris." *British Medical Journal* 1 (April 25): 1065–69.

Prior, I. A., J. M. Stanhope, J. G. Evans, and C. E. Salmond. 1974. "The Tokelau Island Migrant Study." *International Journal of Epidemiology* 3 (Sept.): 225–32.

Prior, I. A., T. J. Welby, T. Østbye, C. E. Salmond, and Y. M. Stokes. 1987. "Migration and Gout: The Tokelau Island Migrant Study." *British Medical Journal* 295 (Aug. 22): 457–61.

Proctor, R. N. 2011. *Golden Holocaust:*

Origins of the Cigarette Catastrophe and the Case for Abolition. Berkeley: University of California Press.
Public Relations Society of America (PRSA). 1976. "The Sugar Association Inc." Campaign profile, two-page summary of a Silver Anvil Award Winner, addressing research, planning, execution and evaluation. No. 6BW-7604C.
Putnam, J. J., and S. Haley. 2003. "Estimating Consumption of Caloric Sweeteners." *Amber Waves* (April 1). http://www.ers.usda.gov/amber-waves/2003-april/behind-the-data.aspx#.V1meFpMrJXs.
Quinzio, J. 2009. *Of Sugar and Snow: A History of Ice Cream Making.* Berkeley: University of California Press.
Reader, G., R. Melchionna, L. E. Hinkle, et al. 1952. "Treatment of Obesity." *American Journal of Medicine* 13, no. 4 (Oct.): 478–86.
Reaven, G. M. 1997. "The Kidney: An Unwilling Accomplice in Syndrome X." *American Journal of Kidney Diseases* 30, no. 6 (Dec.): 928–31.
———. 1988. "Banting Lecture 1988: Role of Insulin Resistance in Human Disease." *Diabetes* 37, no. 12 (Dec.): 1595–1607.
Reed, A. C. 1916. "Diabetes in China." *American Journal of the Medical Sciences*

151, no. 4 (April): 577–81.
Reiser, S. 1987. "Uric Acid and Lactic Acid." In Reiser and Hallfrisch 1987, 113–34.
Reiser, S., and J. Hallfrisch. 1987. *Metabolic Effects of Dietary Fructose.* Boca Raton, Fla.: CRC Press.
Reiser, S., J. Hallfrisch, J. Fields, et al. 1986. "Effects of Sugars on Indices of Glucose Tolerance in Humans." *American Journal of Clinical Nutrition* 43, no. 1 (Jan.): 151–59.
Reiser, S., and B. Szepesi. 1978. "SCOGS Report on the Health Aspects of Sucrose Consumption." *American Journal of Clinical Nutrition* 31, no. 1 (Jan.): 9–11.
Review Panel of the National Heart Institute. 1969. "Mass Field Trials of the Diet-Heart Question, Their Significance, Feasibility and Applicability: Report of the Diet-Heart Review Panel of the National Heart Institute." American Heart Association Monograph no. 28. American Heart Association.
Reynolds, G. 2014. "Drink Soda? Take 12,000 Steps." *New York Times.* Sept. 10. At http://well.blogs.nytimes.com/2014/09/10/drink-soda-keep-walking/?_php=true&_type=blogs& ref=health&_r=0.

Rhein, R. W., Jr., and L. Marion. 1977. *The Saccharin Controversy: A Guide for Consumers.* New York: Monarch Press.

Richardson, T. 2002. *Sweets: A History of Candy.* New York: Bloomsbury.

Rippe, J. M., and T. J. Angelopoulos. 2015. "Fructose-Containing Sugars and Cardiovascular Disease." *Advances in Nutrition* 6, no. 4 (July 15): 430–39.

Ripperger, H. 1934. "America's Huge Appetite for Candy." *New York Times,* July 15.

Roberts, A. M. 1973. "Effects of a Sucrose-Free Diet on the Serum-Lipid Levels of Men in Antarctica." *Lancet* 301 (June 2): 1201–4.

Rollo, J. 1798. "The History, Nature and Treatment of Diabetes Mellitus." In *Diabetes: A Medical Odyssey* (Tuckahoe, N.Y.: USV Pharmaceutical Corp., 1971), 23–44.

Rony, H. R. 1940. *Obesity and Leanness.* Philadelphia: Lea & Febiger.

Root, W., and R. de Rochemont. 1976. *Eating in America: A History.* New York: Ecco Press.

Rose, B. S. 1975. "Gout in the Maoris." *Seminars in Arthritis and Rheumatism* 5, no. 2 (Nov.): 121–45.

Rose, M. S. 1929. *The Foundations of Nutrition.* New York: Macmillan.

Rosenberg, C. E. 1987. *The Care of Strangers: The Rise of America's Hospital System.* Baltimore: Johns Hopkins University Press.

Rosenthal, B., M. Jacobson, and M. Bohm. 1976. "Professors on the Take." *Progressive,* Nov.: 42–47.

Rush, E. and L. Pearce. 2013. *Foods Imported into the Tokelau Islands: 10th May 2008 to 1 April 2012.* World Health Organization (Western Pacific Region). At http://aut.researchgateway.ac.nz/handle/10292/5757.

Russell, F. 1975. *The Pima Indians.* Tucson: University of Arizona Press. [Originally published 1905.]

Sagild, U., J. Littauer, C. S. Jespersen, and S. Andersen. 1966. "Epidemiological Studies in Greenland 1962–1964: 1, Diabetes Mellitus in Eskimos." *Acta Medica Scandinavica* 179, no. 1 (Jan.): 29–39.

Saundby, R. 1908. "Diabetes Mellitus Among the Chinese." *British Medical Journal* 1 (Jan. 11): 116–17.

———. 1901. "Diabetes Mellitus." In *A System of Medicine,* ed. T. C. Allbutt (New

York: Macmillan), 195–233.
———. 1891. *Lectures on Diabetes: Including the Bradshawe Lecture, Delivered Before the Royal College of Physicians on August 18th, 1890.* New York: E. B. Treat.
Schaefer, O. 1968. "Glycosuria and Diabetes Mellitus in Canadian Eskimos." *Canadian Medical Association Journal* 99, no. 5 (Aug. 3): 201–6.
Schmidt, L. A. 2015. "What Are Addictive Substances and Behaviours and How Far Do They Extend?" In *The Impact of Addictive Substances and Behaviours on Individual and Societal Well-Being,* eds. P. Anderson, J. Rehm, and R. Room (Oxford, U.K.: Oxford University Press), 37–52.
Schmitz, A., and D. Christian. 1993. "The Economics and Politics of U.S. Sugar Policy." In Marks and Maskus, eds., 1993, 49–78.
Schneider, J. A., Z. Arvanitakis, W. Bang, and D. A. Bennett. 2007. "Mixed Brain Pathologies Account for Most Dementia Cases in Community-Dwelling Older Persons." *Neurology* 69, no. 24 (Dec. 11): 2197–2204.
Schulz, K. 2010. *Being Wrong: Adventures in the Margin of Error.* New York: HarperCollins.

Schweitzer, A. 1957. "Preface." In A. Berglas, *Cancer: Nature, Cause and Cure* (Paris: Institut Pasteur), ix.

Sclafani, A. 1987. "Carbohydrate, Taste, Appetite, and Obesity: An Overview." *Neuroscience and Biobehavioral Reviews* 11, no. 2 (Summer): 131–53.

Seegmiller, J. E., R. M. Dixon, G. J. Kemp, et al. 1990. "Fructose-Induced Aberration of Metabolism in Familial Gout Identified by 31P Magnetic Resonance Spectroscopy." *Proceedings of the National Academy of Sciences* 87, no. 21 (Nov.): 8326–30.

Seidenberg, C. 2015. "How to Teach Your Kids About Sugar." *Washington Post.* May 13. At https://www.washingtonpost.com/lifestyle/wellness/ how-to-teach-your-kids-about-sugar/2015/05/12/6b8b7882-f401-11e4-b2f3-af5479e6bbdd_story.html.

Select Committee on Nutrition and Human Needs of the U.S. Senate. 1977. *Dietary Goals for the United States.* Washington, D.C.: U.S. Government Printing Office.

———. 1973. *Sugar in Diet, Diabetes, and Heart Disease.* Hearing Before the Select Committee on Nutrition and Human Needs of the United States Senate, 93rd Congress, pt. 2. April 30, May 1 and 2, 1973. Washington, D.C.: U.S. Govern-

ment Printing Office.
Shafrir, E. 1991. "Fructose/Sucrose Metabolism: Its Physiological and Pathological Implications." In Kretchmer and Hollenbeck, eds., 1991, 63–98.
Shaper, A. G. 1967. "Blood Pressure Studies in East Africa." In *The Epidemiology of Hypertension,* ed. J. Stamler, R. Stamler, and T. N. Pullman (New York: Grune & Stratton), 139–49.
Shaper, A. G., P. J. Leonard, K. W. Jones, and M. Jones. 1969. "Environmental Effects on the Body Build, Blood Pressure and Blood Chemistry of Nomadic Warriors Serving in the Army in Kenya." *East African Medical Journal* 46, no. 5 (May): 282–89.
Shattuck, G. C. 1937. "The Possible Significance of Low Blood Pressures Observed in Guatemalans and in Yucatecans." *American Journal of Tropical Medicine* 17, no. 4 (July): 513–37.
Shryock, R. H. 1979. *The Development of Modern Medicine: An Interpretation of the Social and Scientific Factors Involved.* Madison: University of Wisconsin Press.
Silver, S., and J. Bauer. 1931. "Obesity, Constitutional or Endocrine?" *American Journal of the Medical Sciences* 181, no. 1

(Jan.): 769–77.

Sipple, H. L., and K. W. McNutt, eds. 1974. *Sugars in Nutrition.* New York: Academic Press.

Siu, R. G. H., J. F. Borzelleca, C. J. Carr, et al. 1977. "Evaluation of Health Aspects of GRAS Food Ingredients: Lessons Learned and Questions Answered." *Federation Proceedings* 36, no. 11 (Oct.): 2519–62.

Slare, F. 1715. "Vindication of Sugars Against the Charge of Dr. Willis, Other Physicians, and Common Prejudices." In *Observations upon BEZOAR-stones: With a Vindication of Sugars, &c.* London: Tim Goodwin.

Smith, C. J., E. M. Manahan, and S. G. Pablo. 1994. "Food Habit and Cultural Changes Among the Pima Indians." In Joe and Young, eds., 1994, 407–33.

Smith, D. 1952. "Fight Continues Between Dentists, Sugar Industry." *Boston Globe,* Sept. 1: 34.

Snapper, I. 1960. *Bedside Medicine.* New York: Grune & Stratton.

Sniderman, A. D., K. Williams, J. H. Contois, et al. 2011. "A Meta-Analysis of Low-Density Lipoprotein Cholesterol, Non-High-Density Lipoprotein Cholesterol, and Apolipoprotein B as Markers of Cardiovascular Risk." *Circulation: Cardiovascular Quality and Outcomes* 4, no. 3 (May): 337–45.

Snowden, C. 2015. "The Coca-Cola 'Exposé' Had All the Spin of a Classic Anti-Sugar Smear Piece." *Spectator,* Oct. 12. At https://health.spectator.co.uk/the-coca-cola-expose-had-all-the-spin-of-a-classic-anti-sugar-smear-piece/.

Snowdon, D. A., L. H. Greiner, J. A. Mortimer, K. P. Riley, P. A. Greiner, and W. R. Markesbery. 1997. "Brain Infarction and the Clinical Expression of Alzheimer's Disease: The Nun Study." *J.A.M.A.* 277, no. 10 (March 12): 813–17.

Sorem, K. A. 1985. "Cancer Incidence in the Zuni Indians of New Mexico." *Yale Journal of Biology and Medicine* 58, no. 5 (Sept.–Oct.): 489–96.

Standage, T. 2005. *A History of the World in 6 Glasses.* New York: Walker & Company.

Stare, F. J. 1987. *Harvard's Department of Nutrition 1942–1986.* Norwell, Mass.: Christopher Publishing House.

———. 1976a. "The Consequences of Reducing Sugar." *Trends in Biochemical Sciences* 1, no. 10 (Oct.): 226.

———. 1976b. "Sugar Is a Cheap Safe Food." *Trends in Biochemical Sciences* 1,

no. 6 (June): N126–28.
———. ed. 1975. “Sugar in the Diet of Man.” *World Review of Nutrition and Dietetics* 22: 237–326.
Starling, S. 2009. “Groups Unite to Fight US Obesity.” Food Navigator–USA.com. October 5. At http://www.foodnavigator-usa.com/Suppliers2/Groups-unite-to-fight-US-obesity.
Steiner, J. E. 1977. “Facial Expressions of the Neonate Infant Indicating the Hedonics of Food-Related Chemical Stimuli.” In *Taste and Development: The Genesis of Sweet Preference,* ed. J. M. Weiffenbach (Washington, D.C.: U.S. Government Printing Office), 173–88.
Stockard, C. R. 1929. “Hormones of the Sex Glands–What They Mean for Growth and Development.” In *Chemistry in Medicine,* ed. J. Stieglitz (New York: Chemical Foundation, Inc., 1929), 256–71.
Stoddard, B. 1997. *Pepsi: 100 Years.* Los Angeles: General Publishing Group.
Strom, S. 2012. “Nation’s Sweet Tooth Shrinks.” *New York Times,* Oct. 27: B1.
Strong, L. A. G. 1954. *The Story of Sugar.* London: Weidenfeld & Nicolson.
Suddick, R. P., and N. O. Harris. 1990. “Historical Perspectives of Oral Biology: A Series.” *Critical Reviews in Oral Biology*

and Medicine 1, no. 2 (June): 135–51.
Sugar Association, Inc. (SAI). 1978. Sugar Association, Inc., winter meeting of the board of directors, Chicago, Ill., Feb. 9, 1978. Research projects report, Washington, D.C. Sugar Association, Inc., Records of the Great Western Sugar Company, Colorado Agricultural Archive, Colorado State University.
———. 1977a. "President's Report." In Sugar Association, Inc., fall meeting of the board of directors, Palm Springs, Calif., Oct. 13. Sugar Association, Inc., Records of the Great Western Sugar Company, Colorado Agricultural Archive, Colorado State University.
———. 1977b. Annual meeting of the board of directors, Chicago, Ill., May 12. Washington, D.C. Internal document, Sugar Association, Inc., Records of the Great Western Sugar Company, Colorado Agricultural Archive, Colorado State University.
———. 1977c. "Report of the President." In SAI 1977b.
———. 1977d. "Report of the Treasurer." In SAI 1977b.
———. 1977e. "Sugar Is Safe!" Washington, D.C. Sugar Association, Inc., Records of the Great Western Sugar Company, Colo-

rado Agricultural Archive, Colorado State University.

———. 1976. Memo from Jack O'Connell, March 15. Internal document. Sugar Association, Inc., Records of the Great Western Sugar Company, Colorado Agricultural Archive, Colorado State University.

———. 1975a. Transcript of the Sugar Association, Inc. program at the Newspaper Food Editors Conference in Chicago. Oct. 10. Internal document. Washington, D.C. Sugar Association, Inc., Records of the Great Western Sugar Company, Colorado Agricultural Archive, Colorado State University.

———. 1975b. SAI Press Release, "Scientists Dispel Sugar Fears," SAI letter to editor, Sept. 26, 1975, "Sugar in the Diet of Man," summary by Ronald M. Deutsch. Sugar Association, Inc., Records of the Great Western Sugar Company, Colorado Agricultural Archive, Colorado State University.

———. 1975c. Confidential memo to Public Relations Committee, July 17. Washington, D.C. Sugar Association, Inc., Records of the Great Western Sugar Company, Colorado Agricultural Archive, Colorado State University.

———. 1975d. Minutes of meeting of Public Communications Committee. April 21. Chicago, Ill., Internal document, Washington, D.C. Sugar Association, Inc., Records of the Great Western Sugar Company, Colorado Agricultural Archive, Colorado State University.

Sugar Information, Inc. 1957. "How Sugar Can Help You Reduce — and Stay There." *Washington Post,* June 18: B4.

———. 1956. "A Timely Report on . . . the Importance of Sugar." *Washington Post,* April 10: 10.

Sugar Institute. 1931a. "Iced Tea! . . . Iced Coffee! . . . Lemonade! . . ." *Boston Globe,* July 30: 25.

———. 1931b. "It's Very Easy to Catch Cold When You Are Tired Out." *Boston Globe,* Feb. 26: 15.

———. 1930. "If You're Tired at 4 o'Clock Get Something to Eat That's Sweet." *Boston Globe,* Oct. 16: 14.

Sugar Research Foundation, Inc (SRF). 1945. *Some Facts About the Sugar Research Foundation, Inc., and Its Prize Award Program.* Oct. Washington: Sugar Research Foundation, Inc.

Sugarman, J. R., M. Hickey, T. Hall, and D. Gohdes. 1990. "The Changing Epidemiology of Diabetes Mellitus Among

Navajo Indians." *Western Journal of Medicine* 153, no. 2 (Aug.): 140–45.

Sugarman, J. R., L. L. White, and T. J. Gilbert. 1990. "Evidence for a Secular Change in Obesity, Height, and Weight Among Navajo Indian Schoolchildren." *American Journal of Clinical Nutrition* 52, no. 6 (Dec.): 960–66.

Swift, T. P. 1937. "Battle on Sugar Dates to War Days." *New York Times,* Sept. 5: 31.

Symonds, B. 1923. "The Blood Pressure of Healthy Men and Women." *J.A.M.A.* 80, no. 4 (Jan. 27): 232–36.

Szanto, S., and J. Yudkin. 1969. "The Effect of Dietary Sucrose on Blood Lipids, Serum Insulin, Platelet Adhesiveness and Body Weight in Human Volunteers." *Postgraduate Medical Journal* 45 (Sept.): 602–7.

Talhout, R., A. Opperhuizen, and J. G. van Amsterdam. 2006. "Sugars as Tobacco Ingredient: Effects on Mainstream Smoke Composition." *Food and Chemical Toxicology* 44, no. 11 (Nov.): 1789–98.

Tanchou, S. 1844. *Recherches sur le traitement médical des tumeurs cancéreuses.* Paris: Gerner Baillière.

Tappy, L., and E. Jéquier. 1993. "Fructose and Dietary Thermogenesis." Supplement,

American Journal of Clinical Nutrition 58, no. 5 (Nov.): S766–S770.

Tappy, L., and L.-A. Lê. 2010. "Metabolic Effects of Fructose and Worldwide Increase in Obesity." *Physiological Reviews* 90, no. 1 (Jan.): 23–46.

Tatem J. W., Jr., 1976a. "President's Report." In board of directors meeting. Oct. 14, Internal document, Scottsdale, Ariz. Sugar Association, Inc., Records of the Great Western Sugar Company, Colorado Agricultural Archive, Colorado State University.

———. 1976b. Letter to Lewis Bergman, editor of *The New York Times Magazine*. June 25. Sugar Association, Inc., Records of the Great Western Sugar Company, Colorado Agricultural Archive, Colorado State University.

———. 1976c. "Remarks: John W. Tatem, Jr., President, The Sugar Association, Inc., to the Chicago Nutrition Association Symposium on Sugar in Nutrition." Jan. 19. Internal document. Sugar Association, Inc., Records of the Great Western Sugar Company, Colorado Agricultural Archive, Colorado State University.

———. 1975. "Status of Sweeteners in the USA: Remarks by John Tatem, International Sugar Meetings, Paris, France."

Nov. 27. Internal document, Washington, D.C. Sugar Association, Inc., Records of the Great Western Sugar Company, Colorado Agricultural Archive, Colorado State University.

Tattersall, R. 2009. *Diabetes: The Biography.* Oxford, U.K.: Oxford University Press.

Taubes, G. 2012. "Cancer Research: Unraveling the Obesity-Cancer Connection." *Science* 335 (Jan. 6): 28–32.

———. 2009. "Insulin Resistance: Prosperity's Plague." *Science* 325 (July 17): 256–60.

———. 2007. *Good Calories, Bad Calories.* New York: Knopf.

Temin, H. M. 1968. "Carcinogenesis by Avian Sarcoma Viruses: X, The Decreased Requirement for Insulin-Replaceable Activity in Serum for Cell Multiplication." *International Journal of Cancer* 3, no. 6 (Nov. 15): 771–87.

———. 1967. "Studies on Carcinogenesis by Avian Sarcoma Viruses: VI, Differential Multiplication of Uninfected and of Converted Cells in Response to Insulin." *Journal of Cell Physiology* 69, no. 3 (June): 377–84.

Thomas, D. B. 1979. "Epidemiologic Studies of Cancer in Minority Groups in the Western United States." *National Cancer*

Institute Monograph no. 53 (Nov.): 103–13.

Thomas, L. 1985. "Medicine as a Very Old Profession." In *Cecil Textbook of Medicine,* 17th edition, ed. J. B. Wyngaarden and L. H. Smith, Jr. (Philadelphia: W. B. Saunders), 9–11.

Thomas, W. A. 1928. "Health of a Carnivorous Race: A Study of the Eskimo." *J.A.M.A.* 88, no. 20 (May 14): 1559–60.

Thorne, V. B. 1914. "Effects of Different Foods upon the Growth of Cancer." *New York Times,* March 1: SM6.

Tilley, N. M. 1972. *The Bright Tobacco Industry: 1860–1929.* New York: Arno Press.

Timberlake, C. 1983. "Diet Soft Drinks Becoming Heavyweight in U.S. Market." *Reading Eagle,* Feb. 20: 52.

Today show. 1976. April 8. Television program. At http://www.nbcuniversal archives .com/nbcuni/clip/5112796793_s02.do.

Trowell, H. C. 1981. "Hypertension, Obesity, Diabetes Mellitus and Coronary Heart Disease." In Trowell and Burkitt, eds., 1981, 3–32.

———. 1975. "Obesity in the Western World." *Plant Foods for Man* 1: 157–68.

———. 1947. "A Case of Gout in a Ruanda, African." *East African Medical Jour-*

nal, 24 (Oct.): 346–48.
Trowell, H. C., and D. P. Burkitt. 1981. "Preface." In Trowell and Burkitt, eds., 1981, xiii–xvi.
Trowell, H. C., and D. P. Burkitt, eds. 1981. *Western Diseases: Their Emergence and Prevention.* Cambridge, Mass.: Harvard University Press.
Trowell, H. C., and S. A. Singh. 1956. "A Case of Coronary Heart Trouble in an African." *East African Medical Journal* 33: 391–94.
Truswell, A. S. 1977. "Dietary Fat and Heart Disease." *Lancet* 310 (Dec. 3): 1173.
Tuia, I. 2001. "The Tokelau Connection." In *The Health of Pacific Societies: Ian Prior's Life and Work,* ed. P. Howden-Chapman and A. Woodward (Aoteroa, N.Z.: Steele Roberts, 2001), 32–39.
Twain, M. 2010. *Autobiography of Mark Twain: The Complete and Authoritative Edition.* Vol. 1. Berkeley: University of California Press.
Umegaki, H. 2014. "Type 2 Diabetes as a Risk Factor for Cognitive Impairment: Current Insights." *Clinical Interventions in Aging* 9 (June 28): 1110–19.
Urbinati, G.C. 1975. "Hillebrand SS. (Ed.): Is the Risk of Becoming Diabetic Affected

by Sugar Consumption." *Acta Diabetologica Latina* 12, nos. 3–4 (May–Aug.): 256–57.

U.S. Congress. 1958. *Food Additive Amendment of 1958, Public Law.* Sept. 6. 85th Congress, 72 Stat. 1784–89, Washington, D.C.

U.S. Department of Agriculture (USDA). 2016. "Loss-Adjusted Food Availability Documentation. http://www.ers.usda.gov/data-products/food-availability-(per-capita)-data-system/loss-adjusted-food-availability-documentation.aspx#.UZ0clitATH0.

———. 1942. "For Health . . . Eat Some Food from Each Group . . . Every Day!" At http://www.todayifoundout.com/wp-content/uploads/2013/09/The-Basic-Seven.jpg.

U.S. Department of Agriculture (USDA) and U.S. Department of Health, Education, and Welfare (HEW). 1985. "Nutrition and Your Health: Dietary Guidelines for Americans," 2nd edition. Home and Garden Bulletin no. 232. Washington, D.C.

———. 1980. "Nutrition and Your Health: Dietary Guidelines for Americans." Home and Garden Bulletin, no. 228.

U.S. Department of Health and Human

Services (US HHS). 1988. *The Surgeon General's Report on Nutrition and Health.* Washington, D.C.: U.S. Government Printing Office.

U.S. Department of Health, Education, and Welfare (US HEW). 1971. *Arteriosclerosis: A Report by the National Heart and Lung Institute Task Force on Arteriosclerosis.* 2 vols. U.S. Department of Health, Education, and Welfare publication nos. (NIH) 72-137 and 72-219. Washington, D.C.: National Institutes of Health.

U.S. Food and Drug Administration (FDA). 2015. "History of the GRAS List and SCOGS Reviews." At http://www.fda.gov/Food/IngredientsPackagingLabeling/GRAS/SCOGS/ucm084142.htm.

———. 1958. "Food Additives." *Federal Register.* December 9: 23 (239): 9511–17.

Vander Heiden, M. G., L. C. Cantley, and C. B. Thompson. 2010. "Understanding the Warburg Effect: The Metabolic Requirements of Cell Proliferation." *Science* 324 (May 22): 1029–33.

Vaughan, J. 1818. "Abstract and Results from Eight Annual Statements (1809 to 1816), Published by the Board of Health, of the Deaths, with the Diseases, Ages, &c. in the City and Liberties of Philadel-

phia." *Transactions of the American Philosophical Society* 1: 430–34, 453–54.
Ventura, E. E., J. N. Davis, and M. I. Goran. 2011. "Sugar Content of Popular Sweetened Beverages Based on Objective Laboratory Analysis: Focus on Fructose Content." *Obesity* 19, no. 4 (April): 868–74.
Vermeer, S. E., N. D. Prins, T. den Heijer, A. Hofman, P. J. Koudstaal, and M. M. Breteler. 2003. "Silent Brain Infarcts and the Risk of Dementia and Cognitive Decline." *New England Journal of Medicine* 348, no. 13 (March 27): 1215–22.
Veterans Health Administration (VHA). 2011. "Close to 25 Percent of VA Patients Have Diabetes." At http://www.va.gov/health/NewsFeatures/20111115a.asp.
Vinik, A. I., P. A. Crapo, S. J. Brink, et al. 1987. "Nutritional Recommendations and Principles for Individuals with Diabetes Mellitus: 1986." *Diabetes Care* 10, no. 1 (Jan.–Feb.): 126–32.
Walker, G. 1959. "The Great American Dieting Neurosis." *New York Times*, Aug. 23: SM12.
Walter, B. J. 1974. "Sweetener Economics." In *Symposium: Sweeteners*, ed. G. E. Inglett (Westport, Conn.: Avi Publishing, 1974), 45–62.

Walvin, J. 1997. *Fruits of Empire: Exotic Produce and British Taste, 1660–1800.* New York: New York University Press.
Warfield, L. M. 1920. *Arteriosclerosis and Hypertension,* 3rd edition. Saint Louis: C. V. Mosby.
Warner, D. J. 2011. *Sweet Stuff: An American History of Sweeteners from Sugar to Sucralose.* Washington, D.C.: Smithsonian Institution Scholarly Press.
Warren, J. L. 1972. "Sugar — The Question Is, Do We Need It at All? *New York Times,* July 4: 36.
Weidman, D. 2012. "Native American Embodiment of the Chronicities of Modernity." *Medical Anthropology Quarterly* 26, no. 4 (Dec.): 595–612.
Weiss, F. J. 1950. "Tobacco and Sugar." Sugar Research Foundation Inc. Oct. Member report no. 22. At https://www.industrydocumentslibrary.ucsf.edu/tobacco/docs/mjdm0101.
Wells, J. 2014. "Sugar v. Corn Syrup: Sweeteners Clash in Court." CNBC. Jan. 23. At http://www.cnbc.com/2014/01/23/legal-fight-between-sugar-and-corn-syrup-groups-rages-on.html.
Welsh, J. A., S. Karpen, and M. B. Vos. 2013. "Increasing Prevalence of Nonalcoholic Fatty Liver Diseases Among United

States Adolescents, 1988–1994 to 2007–2010." *Journal of Pediatrics* 162, no. 3 (March): 496–500.

Wessen, A. 2001. "Ian Prior and the Tokelau Island Migrant Studies." In *The Health of Pacific Societies: Ian Prior's Life and Work,* ed. P. Howden-Chapman and A. Woodward (Aoteroa, N.Z.: Steele Roberts, 2001), 16–25.

Wessen, A. F., A. Hooper, J. Huntsman, I. A. Prior, and C. E. Salmond, eds. 1992. *Migration and Health in a Small Society: The Case of Tokelau.* Oxford, U.K.: Clarendon Press.

West, K. M. 1978. *Epidemiology of Diabetes and Its Vascular Lesions.* New York: Elsevier.

———. 1974. "Diabetes in American Indians and Other Native Populations of the New World." *Diabetes* 23, no. 10 (Oct.): 841–55.

Whelan, E. M., and F. J. Stare. 1983. *The One-Hundred-Percent Natural, Purely Organic, Cholesterol-Free, Megavitamin, Low-Carbohydrate Nutrition Hoax.* New York: Atheneum.

White, P., and E. P. Joslin. 1959. "The Etiology and Prevention of Diabetes." In *The Treatment of Diabetes Mellitus,* 10th edition, ed. E. P. Joslin, H. F. Root, P. White, and A. Marble (Philadelphia: Lea & Febiger, 1959), 47–98.

White, W. S. 1945. "House Group Warns of Crisis in Sugar." *New York Times,* May 22: 21.

Whitehouse, F. W., and W. J. Cleary, Jr. 1966. "Diabetes Mellitus in Patients with Gout." *J.A.M.A.* 197, no. 2 (July 11): 113–16.

Wilde, O. 1908. *The Picture of Dorian Gray.* Leipzig: Bernhard Tauchnitz.

Wilder, R. M. 1940. *Clinical Diabetes Mellitus and Hyperinsulinism.* Philadelphia: W. B. Saunders.

Wilder, R. M., and D. L. Wilbur, 1938. "Diseases of Metabolism and Nutrition." *Archives of Internal Medicine* 61, no. 2 (Feb.): 297–365.

Willaman, J. J. 1928. "The Race for Sweetness." *Scientific Monthly* 26, no. 1 (Jan.): 76–78.

Williams, R. H., W. H. Daughaday, W. F. Rogers, S. P. Asper, and B. T. Towery. 1948. "Obesity and Its Treatment, with Particular Reference to the Use of Anorexigenic Compounds." *Annals of Internal Medicine* 29, no. 3 (Sept. 1): 510–32.

Williams, W. R. 1945. "Shortage of Sugar

Expected to Last." *New York Times,* May 13: 54.

Willis, T. 1685. *The London Practice of Physick, Or the Whole Practical Part of Physick Contained in the Works of Dr. Willis.* London: Thomas Baffet.

———. 1679. "Of the Diabetes or Pissing Evil." In *Diabetes: A Medical Odyssey* (Tuckahoe, N.Y.: USV Pharmaceutical Corp., 1971), 7–22.

Woloson, W. A. 2002. *Refined Tastes: Sugar, Confectionary, and Consumers in Nineteenth-Century America.* Baltimore: Johns Hopkins University Press.

World Cancer Research Fund (WCRF) and American Institute for Cancer Research (AICR). 1997. *Food, Nutrition and the Prevention of Cancer: A Global Perspective.* Washington, D.C.: American Institute for Cancer Research.

World Health Organization (WHO). 2015. "Obesity and Overweight." At http://www.who.int/mediacentre/factsheets/fs311/en/.

Wyngaarden, J. B., and W. N. Kelley, eds. 1976. *Gout and Hyperuricemia.* New York: Grune & Stratton.

Xu, Y., L. Wang, J. He, et al. 2013. "Prevalence and Control of Diabetes in Chinese Adults." *J.A.M.A.* 310, no. 9 (Sept. 4): 948–58.

Yalow, R. S., and S. A. Berson. 1960. "Immunoassay of Endogenous Plasma Insulin in Man." *Journal of Clinical Investigation* 38, no 7 (July 1): 1157–75.

Yatabe, M. S., J. Yatabe, M. Yoneda, et al. 2010. "Salt Sensitivity Is Associated with Insulin Resistance, Sympathetic Overactivity, and Decreased Suppression of Circulating Renin Activity in Lean Patients with Essential Hypertension." *American Journal of Clinical Nutrition* 92, no. 1 (July): 77–82.

Yoshitake, T., Y. Kiyohara, I. Kato, et al. 1995. "Incidence and Risk Factors of Vascular Dementia and Alzheimer's Disease in a Defined Elderly Japanese Population: The Hisayama Study." *Neurology* 45, no. 6 (June): 1161–68.

Young, T. K., J. Reading, B. Elias, and J. D. O'Neil. 2000. "Type 2 Diabetes Mellitus in Canada's First Nations: Status of an Epidemic in Progress." *Canadian Medical Association Journal* 163, no. 5 (Sept. 5): 561–66.

Yudkin, J. 1986. *Pure, White and Deadly.* Revised edition. New York: Viking.

———. 1972a. *Pure, White and Deadly.* London: Davis-Poynter.

———. 1972b. *Sweet and Dangerous.* New York: P. H. Wyden.

———. 1971. "Sucrose in the Aetiology of Coronary Thrombosis and Other Diseases. In Yudkin, Edelman, and Hough, eds. 1971, 232–41.
———. 1963. "Nutrition and Palatability with Special Reference to Myocardial Infarction, and Other Diseases of Civilization." *Lancet* 281 (June 22): 1335–38.
———. 1957. "Diet and Coronary Thrombosis: Hypothesis and Fact." *Lancet* 273 (July 27): 155–62.
Yudkin, J., J. Edelman, and L. Hough, eds. 1971. *Sugar.* London: Butterworths.
Yudkin, J., V. V. Kakkar, and S. Szanto. 1969 "Sugar Intake, Serum Insulin and Platelet Adhesiveness in Men With and Without Peripheral Vascular Disease." *Postgraduate Medical Journal* 45 (Sept.): 608–11.
Zelman, S. 1950. "The Liver in Obesity." *Annals of Internal Medicine* 90, no. 2 (Aug.): 141–56.
Zhu, Y., B. J. Pandya, and H. K. Choi. 2011. "Prevalence of Gout and Hyperuricemia in the U.S. General Population." *Arthritis and Rheumatism* 63, no. 10 (Oct.): 3136–42.
Ziegler, R. G., R. N. Hoover, M. C. Pike, et al. 1993. "Migration Patterns and Breast Cancer Risk in Asian-American Women." *Journal of the National Cancer Institute* 85,

no. 22 (Nov. 17): 1819–27.
Zimmet, P., K. G. Alberti, and J. Shaw. 2001. "Global and Societal Implications of the Diabetes Epidemic." *Nature* 414 (Dec. 13): 782–87.

INDEX

ABOUT THE AUTHOR

Gary Taubes is the author of *Why We Get Fat* and *Good Calories, Bad Calories.* He is a former staff writer for *Discover* and a correspondent for the journal *Science.* His writing has appeared in *The New York Times Magazine, The Atlantic,* and *Esquire,* and has been included in numerous "Best of" anthologies, including *The Best of the Best American Science Writing* (2010). He has received three Science in Society Journalism Awards from the National Association of Science Writers. He is the recipient of a Robert Wood Johnson Foundation Investigator Award in Health Policy Research and a cofounder of the Nutrition Science Initiative (NuSI). He lives in Oakland, California.

The employees of Thorndike Press hope you have enjoyed this Large Print book. All our Thorndike, Wheeler, and Kennebec Large Print titles are designed for easy reading, and all our books are made to last. Other Thorndike Press Large Print books are available at your library, through selected bookstores, or directly from us.

For information about titles, please call:

(800) 223-1244

or visit our Web site at:

http://gale.cengage.com/thorndike

To share your comments, please write:

Publisher
Thorndike Press
10 Water St., Suite 310
Waterville, ME 04901